Deserts of America

MANAGING EDITORS
Amy Bauman
Barbara J. Behm

CONTENT EDITORS
Amanda Barrickman
James I. Clark
Patricia Lantier
Charles P. Milne, Jr.
Katherine C. Noonan
Christine Snyder
Gary Turbak
William M. Vogt
Denise A. Wenger
Harold L. Willis
John Wolf

ASSISTANT EDITORS
Ann Angel
Michelle Dambeck
Barbara Murray
Renee Prink
Andrea J. Schneider

INDEXER
James I. Clark

ART/PRODUCTION
Suzanne Beck, Art Director
Andrew Rupniewski, Production Manager
Eileen Rickey, Typesetter

Copyright © 1992 Steck-Vaughn Company

Copyright © 1989 Raintree Publishers Limited Partnership for the English language edition.

Original text, photographs and illustrations copyright © 1985 Edizioni Vinicio de Lorentiis/Debate-Itaca.

All rights reserved. No part of the material protected by this copyright may be reproduced or utilized in any form by any means, electronic or mechanical, including photocopying, recording, or by any information storage and retrieval system, without permission in writing from the copyright owner. Requests for permission to make copies of any part of the work should be mailed to: Copyright Permissions, Steck-Vaughn Company, P.O. Box 26015, Austin, TX 78755. Printed in the United States of America.

Library of Congress Number: 88-18337

2 3 4 5 6 7 8 9 0   97 96 95 94 93 92

Library of Congress Cataloging-in-Publication Data

Wingfield, John C., 1948-
  [Deserti d'America. English]
  Deserts of America / John C. Wingfield.

  — (World nature encyclopedia)
  Translation of: Deserti d'America.
  Includes index.
  Summary: Desert ecology—United States—Juvenile literature. 2. Deserts—United States—Juvenile literature.
  I. Title. II. Series: Natura nel mondo. English.
  QH104.W5613 1988   574.5′2652′0973—dc19   88-18337
  ISBN 0-8172-3325-3

WORLD NATURE ENCYCLOPEDIA

Deserts of America

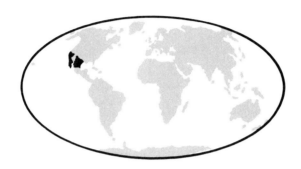

John C. Wingfield

Laramie Junior High
1355 N 22nd
Laramie, WY 82070

RAINTREE
STECK-VAUGHN
LIBRARY

Austin, Texas

CONTENTS

7 INTRODUCTION

9 NORTH AMERICAN DESERTS
The Arid Regions of the American Southwest, 9. The Low Deserts, 11. The Chihuahuan Desert and the Sierra Madre, 15. High Deserts, 19. The Grand Canyon of the Colorado River, 21. The Grand Canyon, Yesterday and Today, 23. Areas North of the Grand Canyon, 26.

29 VEGETATION OF THE LOW DESERTS
Cacti, 29. Trees and Shrubs, 33.

39 SMALL ANIMALS OF THE LOW DESERTS
The Chemical Defenses of Insects, 39. Eight-legged Poisoners and Biters, 39. Reptiles, 44. Snakes, 47. The Desert Tortoise, 49.

51 BIRDS OF THE LOW DESERT
Wrens and Thrashers, 51. Doves, Quails, and Roadrunners, 51. Dwellers in Cactus, 52. Small Perching Birds, 55.

57 MAMMALS OF THE LOW DESERT
Life of the Kangaroo Rat, 57. Other Desert Rodents, 57. The Desert Shrew, 60. The Peccary, 61. The Kit Fox, 63.

65 THE HIGH DESERT
The Pronghorn, 65. Deer, 66. Jackrabbits, 67. Carnivores, 70. Birds, 74. Amphibians, 75.

79 LIFE IN THE CANYONS
Vegetation, 79. Fish, 80. Amphibians and Reptiles, 82.

87 BIRDS OF THE RIPARIAN FORESTS
Bird of the Trees and Meadows, 87. Daytime Birds of Prey, 87. Other Birds, 89. The Wonderful Hummingbirds, 89. Nighttime Birds of Prey, 93.

95 MAMMALS IN THE CANYONS
Bighorn Sheep, 95. Squirrels and Chipmunks, 95. Small Carnivores, 96. The Cats, 99.

105 GUIDE TO AREAS OF NATURAL INTEREST
United States, 107. Mexico, 122.

123 GLOSSARY

126 INDEX

INTRODUCTION

To most people, the word *desert* brings to mind a never-ending expanse of sand, without water and life. It is a place where temperatures can reach unbearable heights, almost like an oven. In reality, many desert areas in the world have an incredible number of animals and plants. Although water is not plentiful, it can be found if a person knows where to look.

The deserts of North America are particularly rich in animal and plant life. This is due to the wide variety of habitats provided by this seemingly hostile environment. Arid regions stretch from coastal plains far inland. Others lie within huge mountain ranges, and some rise to high plateaus in the interior Great Basin.

The mountain ranges that cross the desert basins and plateaus of North America receive considerable rainfall. They are covered with conifer forests and, in winter, are white with snow. They thus represent an important water source for the deserts lying below. In spring, meltwater runs off the mountains in streams and sometimes floods large areas of desert for several weeks. At this time, following winter rains and spring thawing, the desert blooms.

The American deserts are unique in a number of ways. Although thought of by many as mere wastelands to be avoided, deserts are actually delicate, vulnerable ecosystems. Most of the high deserts have been ravaged due to

overgrazing by domesticated animals such as cattle and sheep. The delicate balance that exists in these environments is threatened by pollution from smelters and power plants. Strip-mining for coal and copper, a process which involves scraping off the soil and upper layers of rock, is also a threat.

Dams have been built along large rivers to provide water and electricity for the ever-increasing population of the region and for the rich farmlands of the Imperial Valley in southern California. Clearly, the rapidly growing population of the American Southwest needs water and electricity, but the frail desert ecosystem is subject to severe strain in the process. Water trapped behind dams is not sufficient for irrigating fields, lawns, and golf courses, and for providing domestic supplies. So people have resorted to using "fossil water," which is underground water accumulated over thousands—even millions—of years. This water supply is being pumped and exhausted at an alarming rate.

In a sense this is plundering, done with little regard for the future. How much longer will it last? One can only hope that further development in the Southwest will be accompanied by a more enlightened attitude toward the desert ecosystems. If not, the image of the desert as a desolate stretch of land, with no water or life, might actually become reality.

NORTH AMERICAN DESERTS

There are two types of deserts on this planet. The first type is always located along the "horse latitudes." The horse latitudes are either of two bands located at about 30 degrees latitude, characterized by high pressure, light winds, and light rainfall. Deserts of the first type generally occur at about 30 degrees latitude on the western edges of continents. Examples of these are the Sahara and Namib deserts in Africa, the Spanish deserts in Europe, the Atacama in South America, and the great Indian and Australian deserts.

The second type of desert is found in "rain shadows." Deserts caused by rain shadowing can form anywhere on the lee side of mountain ranges. Moist air moving in from the oceans rises as it reaches the mountains and condenses into large clouds. Rain falls all along the seaward slopes before reaching the mountaintops. If the mountain range is large enough, abundant rain will fall on that side, while the lee side receives very little water and remains dry.

The Arid Regions of the American Southwest

The deserts of the American Southwest are a combination of both desert types. In southern California and northern Mexico, deserts that originated due to the effect of latitude are more common. Inland, deserts are found on the lee side of the Sierra Nevadas, a mountain range that runs north to south in California.

Other lofty mountain ranges, like those in Nevada and Utah, also contain arid regions. Desert-forming mountain ranges dot the Southwest regions, where no one stretch of desert is more than 120 to 190 miles (200 to 300 kilometers) across. These deserts form a vast network stretching in a north-south direction for over 1,200 miles (2,000 km) from Oregon and Idaho down to Mexico, and in an east-west direction for over 900 miles (1,500 km) from California to Texas.

Within this huge area, four principal regions are found. The first is the Great Basin region, which stretches from southeastern Oregon and southern Idaho through much of Nevada and Utah down to northern Arizona, southeastern Colorado, and extreme eastern California. Most of the deserts in this region are at high elevations, occurring around 5,000 feet (1,500 m) and above. Most of the precipitation here falls in the winter months as snow. Summers are short. The vegetation consists mainly of shrubs with small leaves. Trees grow only in well-sheltered environments along rivers.

Preceding pages: Monument Valley, with its typical rock buttes jutting out of arid plains, is perhaps the most symbolic image of North American deserts.

Opposite: The Chihuahuan Desert seems to stretch endlessly when observed from the top of the Chisos Mountains. These mountains are on the border between the United States and Mexico.

The drawing shows how deserts are formed as a result of rain-shadowing. Masses of humid air *(blue arrows)* head inland from the ocean. When they reach a mountain range, the air is forced to rise along the mountain slopes. Cooling and condensation at higher altitudes produce rain. The rain allows lush vegetation to develop on the side of the mountain range facing the sea. Beyond the top of the mountains, the air masses become dry and hot. This, in turn, creates desert conditions.

The second region, the Mojave Desert, is limited to southeastern California and a small area in southern Nevada and western Arizona. This desert, which is also relatively high, is found at elevations of 2,600 feet (800 meters) and above. Precipitation falls mainly in winter. Frost is frequent, and as a result, the shrubs are low and stunted. Few trees and succulent plants can survive these harsh conditions.

The third region is the Chihuahuan Desert, covering part of southeastern Arizona, southern New Mexico, southwest Texas, the Mexican state of Chihuahua, and other regions farther to the south. Much of this land is also at high elevations, and winters can be severe. Most rain falls in late summer, and the typical vegetation includes low shrubs, succulent plants, and trees along rivers.

The richest and most varied desert region, however, is probably the fourth one, the great Sonoran Desert. It extends over most of southern Arizona, extreme southeastern California, and the Mexican states of Sonora and Baja California. It is a low desert, extending from sea level up to 650 feet (200 m). Unlike the other deserts, it rarely experiences frost during the winter, and rainfall occurs both in winter and summer. This allows rich plant and animal life to develop, with "forests" of huge cacti, shrubs, and riparian, or river-edge, forests. At higher altitudes, where winters are colder, wide grasslands and oak and juniper forests are found.

The diversity of landscapes in these deserts is astounding. Rocky slopes, flat pans of salt, high canyons, towering cliffs, grassy steppes, and lava flows can all be found. Vast "seas of sand" typical of the Sahara Desert are rare but can

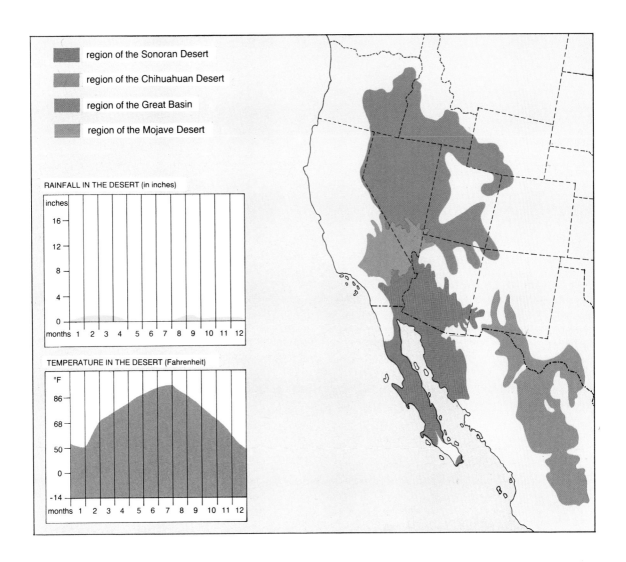

The map shows the distribution of the four principal desert regions in North America. On the side are diagrams for rainfall and temperature in these zones. As shown, North American desert rainfall is always extremely low, and in some months there is no rain at all. The temperature, on the other hand, is always quite high, especially during summer months.

Following pages: The wide expanses of sand and salt that form Death Valley stretch out toward the Sierra Nevada range.

be found in the Dunes area of Nevada, the White Sands region of New Mexico, and El Gran Desierto of Sonora, Mexico, where some dunes are over 490 feet (150 m) high.

Each type of desert has its own unique vegetation, and some animal species live only in particular desert habitats. Other species, however, are much more widely distributed and might be common in two or more different types of desert. In the following paragraphs, various desert regions will be described in detail, but the more widely distributed animals and plants will be described separately.

The Low Deserts

In southern California, Arizona, and northern Mexico (especially Baja California and the coastal region of Sonora)

the desert is mostly low-lying and intensely hot. In the lower Colorado River Valley, summer temperatures can rise higher than 120°F (49°C), and surface temperatures can even approach 200°F (93°C). Most of this area is encompassed by the Sonoran Desert, but in southeastern California, parts of the Mojave and Great Basin deserts dip below sea level in Death Valley. Death Valley is one of the hottest, driest, and lowest areas in the Western Hemisphere. It is 155 miles (250 km) long and about 2 to 20 miles (3 to 32 km) across. It owes its existence to a fault line running the length of the valley. The area sank as the surrounding mountain ranges were rising. The deep valley tends to trap air and this creates a "furnace" effect. Here, the highest temperature in North America was recorded: 134°F (57°C) in the shade. Rainfall, which averages around 10 inches (250 millimeters) a year in the desert, ranges only between 1 and 7 inches (25 to 180 mm) in Death Valley, and in some years there may be no rainfall at all. What little rain there is often evaporates before it reaches the ground.

Much of Death Valley is a rocky desert, but some dunes of granite and sandstone grains can be found. Typically, rock formations are covered with "desert varnish," which gives everything a dark, shiny look. This peculiar coating is formed by iron and manganese oxides that are dissolved from the rocks by rain. The solution spreads over the rock surface during wet weather. When the sun returns, this mineral solution evaporates and produces a varnishlike coating on the rocks.

The lower desert region has a complex geological history. It was uplifted when the Pacific plate slid beneath the North American plate. The same process created the Sierra Nevada range in California and many other mountains in this region. The movements of the two plates generated tremendous pressure and considerable volcanic activity. Faulting and tilting of rock layers caused some regions to sink, forming a series of mountains that alternate with desert basins. In the areas where sinking did not occur, the remaining land formed plateaus.

During the last 20 to 30 million years, both mountains and basins have been continually eroded by the action of water during summer thunderstorms and by wind-blown sand. At higher altitudes during the winter months, water freezes in the cracks of rocks and splits off small fragments or even large chunks of rock. The result is a bewildering array of granite peaks, sandstone ledges, cliffs, and lime-

The lower deserts have a very complex geological history. The slippage of the Pacific plate beneath the North American plate created tremendous pressures and considerable volcanic activity. This activity led to volcanic rock intrusions, craters, and lava flows in the desert areas. An example of such a landscape is the expanse of lava bordered by volcanic cones, which can be seen in this picture taken near Sunset Crater, Arizona.

stone domes that are deeply cut by small canyons called "arroyos," and surrounded by rocky slopes that lead down to wide flats called "playas."

But this is not all. The whole region shows evidence of volcanic activity, with volcanic intrusions (volcanic rocks found among other rock formations), volcanic craters, and lava flows. In some places, erosion has exposed petrified lumps of lava that once plugged the throats of extinct volcanoes. In other areas, volcanic cones stand over 13,100 feet (4,000 m). Sometimes lava flows look so fresh that, at first glance, they seem to have occurred just yesterday.

To the south, along the border with Mexico, lies the Pinacate Desert. It is bordered by the delta of the Colorado River, the Gulf of California, and the Sierra Madre. Most of this area is flat, except for the Pinacate Peak which stands 3,900 feet (1,200 m) tall. It is swept by hot, dry winds. The largest sand sea in North America, El Gran Desierto (the great desert), is found here. To the west stretches the peninsula of Baja California, 745 miles (1,200 km) long; the Tropic of Capricorn crosses its tip.

Like other areas in this region, the peninsula of Baja California was formed by the uplift of a portion of the earth's crust. Tilting and cracking of rock layers occurred and large areas sank during this uplift. The result was the formation of the Gulf of California and the separation of the peninsula from the rest of the continent. Originally, this region was the bed of an ancient sea lying over granite. Today, the mountain slopes are limestone and sandstone, but the core is still granite.

Wind, water, and frost erosion have all left their marks on this land. Moreover, the western coast is exposed to the action of the Pacific Ocean, which has produced spectacular promontories and rocky islets. On the other hand, some small desert islands along the coast are of volcanic origin. Among these are the Channel Islands off southern California. Islands in the Cortez Gulf of California, also called "Sea of Cortez," were once a part of the peninsula, but they were cut off as sea level rose when the last glaciers melted.

In this region, rain usually comes twice a year, in winter and summer. In winter, occasional tropical hurricanes roar up the Pacific coast, bringing moisture with them. Storms can also originate from extensive low-pressure systems that sweep over California and sometimes reach the Sierra Nevada to drop some light rain on the Sonoran Desert. Spring and fall are usually the driest and hottest periods of the year. In summer, though, the region may be swept by thunderstorms caused by pockets of low pressure moving north from the tropics. The storms can be extremely violent but are usually confined to small areas. One area could receive its entire yearly rainfall from a storm while just over the next mountain range, no rain falls at all. So the total rainfall for an area can vary tremendously, from virtually none in some years up to nearly 16 inches (400 mm) in others.

The Chihuahuan Desert and the Sierra Madre

East of the Sonoran Desert is a region that is topograph-

A picturesque view shows Bahia de la Conception, along the coast of Baja California. The Baja peninsula originated from the uplift of part of the earth's crust, followed by the sinking of land where the Gulf of California is now located. It is mostly arid, and its western coasts have been deeply eroded by the Pacific Ocean. The erosion has given rise to small islands, promontories, and cliffs.

ically quite distinct. This is the Chihuahuan Desert, stretching from southeastern Arizona, New Mexico, and Texas south for 800 miles (1,300 km) toward Mexico City and Guadalajara. The region is dominated by two enormous mountain ranges, the Sierra Madre Occidental and the Sierra Madre Oriental.

The Sierra Madre Occidental range is mainly of volcanic origin and resulted from a tremendously violent geological upheaval. The Sierra Madre Oriental range, on the other hand, is a limestone dome that rose from an ancient seabed where limestone had previously been deposited. The limestone formations are folded and deeply cracked, and water erosion of the porous rock has created vast caves. Some of these caves have collapsed, opening up deep abysses. The more recent volcanic intrusions have caused more folding and tilting, creating the complex geological jumble so typical of this region.

The most arid region, most properly defined as desert,

The so-called white sands give their name to this region of New Mexico. Salt and gypsum accumulated to form these stretches of sand. Erosion of rock formations rich in salt and gypsum started about 25,000 years ago. Today the landscape in this area is completely white, almost like an illusion, and even more parched and lifeless than the classical expanses of sand dunes in African deserts.

lies to the north. It is crossed by the Rio Grande flowing out of New Mexico and along the border of Texas and Mexico. This is the so-called Big Bend region. To the north lie the Chisos Mountains, an outpost of the Sierra Madre range.

The vegetation of the entire region is quite similar to that which exists in the Sonoran Desert. Wide grasslands and many cactus species are seen here, but there is no real "cactus forest" like that encountered in the Sonoran region. Rainfall is scarce, generally coming in late summer and autumn, when storms move north from the humid tropical regions.

High Deserts

Moving north from the floor of the Sonoran Desert is a slope that rises to the Mogollon Rim in Arizona. To the south lie the burning deserts, while to the north are cooler high deserts and more moist country, covered with conifer forests.

During the day on the Painted Desert in Arizona, there is almost no trace of life. The heat is so intense that it rises in waves from the desert floor.

A truly spectacular site, located at the western end of the Rim Country, is Oak Creek Canyon. Here are bright red rocks and spectacular cliffs and buttes. These formations vanish a short distance to the south. The canyon walls display five layers of rock formations. At the bottom there are layers of red siltstone about 2,000 feet (600 m) thick. On top of this come 560 feet (170 m) of light brown sandstone. Next is a layer of white sandstone, 230 feet (70 m) thick, covered by 330 feet (100 m) of gray limestone that was originally deposited on the bottom of an ancient sea. The top layers are composed of black lava. Numerous faults and folds add to the general effect; the result is a kaleidoscope of

colors that change dramatically as the sun rises and sets.

There is always water at the bottom of the canyon. It comes from the thawing of snow and from the rain that falls at higher elevations. The river on the canyon floor supports a rich variety of plant and animal life. As one travels up the canyon, wildlife communities change. They are adapted to desert conditions deep in the canyon and to mountain conditions at higher elevations.

Beyond the Mogollon Rim, a high, dry plateau stretches north for over 620 miles (1,000 km). It is crossed by many mountain ranges, which are covered with conifer forests. To the west is another rather high desert, the Mojave, formed by the rain shadow of the Sierra Nevada on the broad Owens Valley in California. This vast region is highly diverse. Here, flat deserts stretch as far as the eye can see, and there are many deep, intricate canyons. In northern Arizona, the Painted Desert lies north of the Mogollon Rim. It offers breathtaking views of rock formations of all colors: red, yellow, white, black, and even green and blue.

The Grand Canyon of the Colorado River

The Grand Canyon, in northwestern Arizona, is the most spectacular of all desert formations in the world. As vast as it is, the words written about it might well fill it. No descriptions or pictures, though, can come close to giving the visitor the same feeling as the actual view experienced in person.

This amazing canyon is 286 miles (460 km) long and nearly a mile (1,600 m) deep. How did it form? During a ten-million-year period while the plateau was rising, the Colorado River was cutting through the rock layers at an incredible rate. The erosion was made easier by the action of solid particles which the huge mass of water carried along. These particles acted like a scraper on the riverbed. Harder rocks were eroded more slowly, while the softer ones, such as sandstone, were cut through so rapidly that almost vertical walls were formed. At the end of the last glaciation ten thousand years ago, torrents of meltwater must have been loaded with great quantities of rubble and undoubtedly cut the canyon even deeper. The side canyons, where tributaries flow toward the Colorado River, are also huge, especially those lying along fault lines.

Because the river has cut so deeply, the sequence of rock layers that can be seen in the Grand Canyon is extraordinary. It is made up of more than twenty different

The Grand Canyon in Arizona is up to a mile (1,600 m) deep, 4 to 18 miles (7 to 30 km) wide at the top, and about 220 miles (350 km) long. This spectacular geological phenomenon has created a landscape that is one of the world's great natural wonders. The Colorado River, which carved it, collects waters from many tributaries that flow down from snowy mountain peaks in Colorado and Utah.

layers of rock: shale, limestone, sandstone, schist, granite, and many others. At the bottom of the canyon, the schists are two billion years old, while those found at the top are 250 million years old. The rock walls deep in the canyon are almost vertical, but near the top wind erosion has added to the original erosion caused by the river. The amount of rock that has been removed is so large that it cannot even be imagined. Most of it has been deposited in the delta of the Colorado River at the northern end of the Gulf of California, but part of it forms the soil of the Imperial Valley in southern California. Today, this area of rich farmland is completely cultivated, and water from the Colorado River is used for irrigation. Actually, so much water is pumped out of the river in its lower course that by the time it reaches the sea in the Gulf of California, it is no more than a trickle.

The great Colorado River has not always been so tame. At the beginning of the century, massive flooding occurred. As a result, the course of the river changed temporarily and

This diagram shows a cross section of the Grand Canyon of the Colorado River. On the right, rock formations and their geological ages are listed. The fossils found within each layer are also indicated. More than twenty rock layers have been exposed by the action of the Colorado River. The rocks at the bottom, which are around two billion years old, are still poorly understood by geologists. The rocks found near the top of the canyon date back to 250 million years ago. That was the time when the first reptiles appeared on earth.

a wide basin was flooded, forming a new lake. This lake, called the Salton Sea, is now completely landlocked with no outlet. With time, as the hot desert sun evaporates the water, it is becoming more and more salty.

The Grand Canyon, Yesterday and Today

Colorado is a Spanish word meaning "red." Such a name might seem inappropriate because the Colorado River has a greenish color today. Actually, the early Spanish explorers called it the "red river" for good reasons. They were impressed with the huge quantity of reddish silt being carried downstream. The recently built Glen Canyon Dam has completely blocked the flow of silt. As a consequence, the water now appears greenish.

The dam also caused many changes in the Grand Canyon ecosystem. Before the dam's construction, water flow was low in autumn and winter but very high in spring due to snowmelt, and in summer because of rains. When the water level increased, the river would flood its banks and form backwater pools and marshes. These were critical environments for many fish, plants, amphibians, and birds. Today, the water flow is strictly controlled, and these wetlands are no longer formed.

Formerly, up to 9.5 million tons of silt a day were

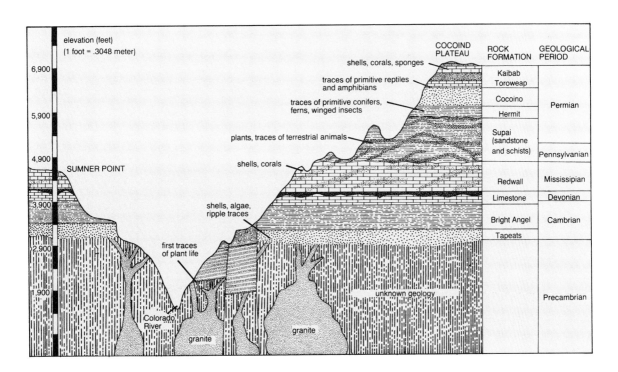

On the right is a view of the Colorado River from the high walls of the Grand Canyon. The river flows into the Gulf of California after crossing eight American states and two Mexican states and covering a distance of 1,550 miles (2,500 km). In the past, this great river experienced major floods. These floods helped erode the relatively soft sandstones beneath a vast plateau. This process created awesome canyons, first and foremost the Grand Canyon. The periodic floods have been eliminated by the construction of the Glen Canyon Dam.

The Colorado River's current is so strong that it can pull fish away from the bottom, where they usually forage for food. For this reason, natural selection has favored some fish species, such as the rare *Gila cypha* (top) and the equally rare razorback sucker, both of which have a hump on their backs. The current actually pushes a fish with such a humped back toward the bottom. Thus the current helps the fish to remain close to the bottom and brings them a constant supply of food.

carried by the river past the river crossing at Lee's Ferry. Today, that amount is only 20 million tons of silt per year. The whole canyon system has been "cleaned up." The water is clearer and sunlight can penetrate the depths. This allows algae to grow rapidly, and they give the river its present color. So today a more appropriate name would be *Rio Verde*, which means "Green River" rather than Rio Colorado.

In the past, the temperature of the river used to fluctuate from a chilly 41°F (5°C) in winter to 86°F (30°C) in the middle of summer. But today, Lake Powell, which is the lake formed behind the Glen Canyon Dam, is so deep that not even the burning desert sun can heat its bottom waters. The turbines of the hydroelectric plants take cold water from the bottom of the lake and release it into the river. Currently, the temperature of the water entering the Grand Canyon is almost constant throughout the year, around 44° to 46°F (7° to 8°C).

Unfortunately, the seasonal changes in temperature were critically important to many aquatic organisms. They provided the signal that triggered the onset of breeding. The loss of this signal has upset the entire ecosystem. Previously, there were eight native fish species in these waters. Today, only three of them survive: the bluehead sucker, the flannelmouth sucker, and the speckled dace. But even these species can only reproduce in the tributaries of the Colorado River, where the water temperature still undergoes seasonal variations, including a warming trend during the summer.

Highly interesting endemic species are thought to be extinct. Endemic species, such as bony-tailed chub, the roundtail chub, and the large squawfish of the Colorado River are those restricted to a certain area. Another endemic species, the razorback sucker, has been reduced to just a small population on the verge of extinction. On the other hand, species such as European carp are rapidly taking over the new environment since being introduced into the river by humans. They feed on the algae, which have also been increasing rapidly in response to the modified habitat.

Not only the river, but the whole bottom of the canyon is undergoing major changes. Control of the water flow has eliminated the scouring and scraping action of floods. As a result, some plants from the rim of the canyon are beginning to grow near the canyon bottom. Some of these plants, such as the tamarisk shrub, have been introduced by people.

The fantastic and desolate landscape of Monument Valley on the Utah-Arizona border is a most appropriate setting for a western movie. The valley is a large, sandy, sunken area, in which only a few animal and plant species can survive. It is famous for its extraordinary spires, chimneys, and buttes, which rise up like towering red fortresses.

Along with the plants, many animals are coming lower into the canyon.

Areas North of the Grand Canyon

North of the Grand Canyon, the land rises to over 8,200 feet (2,500 m). Here are found forests of pines, firs, spruce, and quaking aspen. More to the north and northeast, though, the desert regains its hold. The plains here are studded with huge blocks of harder rock, which form the

landscape of Monument Valley and the mazelike canyons of southern Utah. In these canyonlands, the climate is extreme. Temperatures can fall to -29°F (-34°C) in winter and soar to 109°F (43°C) in summer. Within a day, the temperature can vary as much as 104°F (40°C).

The desert here is definitely rocky, and large stretches of land are without vegetation. Even the cacti are stunted. However, the variety of rock formations makes up for the meager plant life. The harder rocks resist erosion longer, and as a result, monumental buttes are created. These beautiful formations are typical features of the American West. The combined action of water and wind has formed arches and spires that no sculptor could copy. A few coats of "desert varnish" give the final touch to this natural work of art.

Vegetation is sparse in this desert environment. Sagebrush and Mormon tea are the dominant species. Along rivers, oaks, hackberries, dogwoods, and cottonwoods manage to survive, while at higher altitudes pinyon pine and juniper forests are found. In this area, some peaks rise to 14,700 feet (4,500 m), but most of the region is 5,000 to 10,000 feet (1,500 to 3,000 m) high.

Typical features of the Great Basin are salt flats, or playas, and the Great Salt Lake in Utah. Playas are flat expanses of yellow-brown clay, covered with a crust of snow-white salt. When rain comes or snow melts, playas turn into impassable mud holes. Playas have no water drainage whatsoever, since they are completely surrounded by ridges and mountains. The relentless summer sun soon evaporates the water, and the salt recrystallizes and forms a new crust.

The Great Salt Lake is actually the remnant of an ancient glacial lake. It does not have any outlet for its water, so for thousands of years it has been accumulating salt dissolved from the surrounding rocks. This has made the lake's water very salty, currently six times saltier than ocean water.

Despite the high salinity, some organisms manage to survive in this environment. The water is teeming with algae and bacteria, which form pink or turquoise crusts. Some larvae of brine flies feed on these organisms. Pink brine shrimp, just 1/4-inch (6 mm) long, live in large groups in the lake. They are choice food for many bird species that dwell near the lake shore and on the islands.

VEGETATION OF THE LOW DESERTS

The fauna and flora of the low deserts, which have elevations below 3,300 feet (1,000 m), are truly incredible. In these regions, 2,500 plant species have been identified, as well as thousands of invertebrates (animals without backbones) and about five hundred species of vertebrates (animals with backbones). Among the vertebrates are fish, amphibians, reptiles, birds, and mammals. Many species are endemic and are not found elsewhere. All life forms here have adapted to this environment in different ways, but they all must deal with the same problems—how to save water and how to withstand the extreme temperatures.

Cacti

When considering the American deserts, the first plants that come to mind are the cacti. About 140 species of cactus can be found in the deserts of the Southwest. They show a wide variety of shapes but also share some common features. These plants have no leaves. Instead, all cacti are equipped with spines and tough, sharp hairs, and their chlorophyll is contained in stems that have become thick and bulbous. Their outer surface is waxy and helps reduce water loss. Their roots often spread over wide areas just beneath the surface of the soil to absorb as much water as possible when it rains. In fact, water usually does not penetrate very deeply into the desert ground but soon collects to form rivulets and streams. Some cacti, though, also have deep taproots that can reach all the way down to the underground water supply called the "water table." This is particularly useful during dry periods when the only water available may be deep beneath the surface.

A typical landscape in parts of the Sonoran Desert is a forest of giant cacti. The most familiar of these is the saguaro cactus, which can grow as tall as 50 feet (15 m), live up to two hundred years, and weigh many tons when completely filled with water. The saguaro is the giant in its family. Besides all the typical cactus features, this plant has some woody "ribs" to strengthen its inner structure and help support the enormous weight of the water that can be stored. Its skin is pleated so the plant can expand, increasing its circumference up to 50 percent when the spongy inside becomes saturated with water. With a full water supply, the saguaro can easily survive for a whole year without taking in any additional water.

During its first forty years, a saguaro cactus can grow up to 39 feet (12 m), but no branches will emerge before it is

Opposite: Ocotillo bloom in the desert of New Mexico. Despite its rather inconspicuous appearance, this plant is among the species best adapted in dry areas. It can produce leaves and flowers after each rain, even three or four times a year.

seventy years old. It blooms from late May to July, bearing showy, hornlike, yellowish white flowers. The flowers, which are rich in nectar, appear in rosettes at the tip of the main stem and at the end of branches. The nectar is eaten by many insects and birds and even by some reptiles and mammals. The fruit is red and contains a great many seeds. A saguaro can produce up to forty million seeds in its lifetime, of which only three or four will ever grow into an adult plant.

In the southern deserts of Organ Pipe Cactus National Monument in Arizona, and Sonora and Baja California in Mexico, other cactus species are abundant. Typical of these regions are the organ pipe cactus, the cardon cactus, and the senita, or whisker, cactus, all of which are closely related to the saguaro. Some cardon cacti can reach 50 feet (15 m) tall. Their skin and spines resemble those of the saguaro, and they also have woody ribs to support themselves as they become saturated with water. These plants usually bloom at night, but if it is cool enough, the flowers may be seen in the daytime.

The saguaros are the well-known giant cacti of the American deserts. Their blossoms brighten the landscape in this arid region. These huge plants, which can be up to 50 feet (15 m) tall, are a basic resource for desert fauna. In their holes and cavities, they host various birds. Their most frequent guests are the small elf owl and the cactus woodpecker. Many different animals eat saguaro fruit and spread the plant's undigested seeds in their droppings.

Some of the most typical plants of the North American deserts are from left to right *(back row):* saguaro, ocotillo, giant cardon, flowering agave, and teddy-bear cholla; *(front row)* jumping cholla, two specimens of barrel cacti, showing the extreme shapes these plants can take (completely swollen when they are full of water and shrunk when they are almost dehydrated), and two specimens of *Mammillaria cactus.*

Other species of cactus are not as huge as the saguaro and cardon. Some *Mammillaria* cacti resemble clusters of tiny, spiny balls, only 0.8 inch (2 centimeters) in diameter. Species of the genus *Opuntia* include cacti of various shapes and sizes. The best known are the prickly pear and beavertail cacti. Both have a number of flat, branching lobes resembling a beaver's tail with sharp spines and stiff hairs. The yellow or red flowers bloom from the edge of the outer lobes. The seeds are hard, dry, and contained in capsules.

Another group of *Opuntia* is collectively called "chollas." They look quite different from the species already discussed. These shrublike cacti have branching trunks. Numerous spiny lobes project from each branch. Some species can grow to 6 feet (2 m) tall. Perhaps the most beautiful cholla is the so-called teddy bear cholla, which has a thick cover of yellowish spines. Certainly the best-known cholla is the jumping cholla. It owes its name to its loosely attached lobes, which can easily detach and stick to the hair of a passing animal or to the clothing of an unlucky hiker. These lobes even stick to bare skin and have thus gained a reputation for "jumping." These plant parts, which are so easily detached, are actually a means of asexual

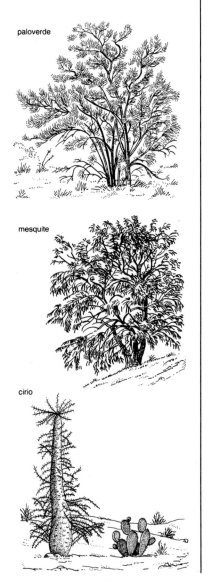

California fan palms—together with mesquite, paloverde, and yucca—are among the most common trees in the American deserts. They have tall trunks, adorned at the top with a tuft of fan-shaped leaves that bend outward and, once dead, stay attached to the trunk.

reproduction. When the lobes eventually fall to the ground, they take root and start a new plant.

Another form of *Opuntia* is known as the "creeping devil" cactus. It can spread on the ground, forming a mat 20 feet (6 m) wide and 1 foot (30 cm) tall. Because of its barbed spines, this cactus can create a formidable barrier to animals and humans.

Another important group of cacti in the lower desert includes the barrel cacti. These are typically columnar cacti; they never form branches. Vertical pleats allow them to expand when they take in water. The main stem is covered with long, sharp spines.

Trees and Shrubs

The low deserts, especially those in the South, almost never experience winter frosts. Therefore, a larger variety of plant species, including many kinds of trees and large shrubs, grows there. These tend to form open forests. Most of the shrubby trees are in the legume, or pea, family. One particularly eye-catching species, which has come to be the trademark of the Sonoran Desert, is the paloverde. This elegant tree can grow to be 26 feet (8 m) high and is armed with sharp thorns at the tip of each twig. It has tiny, bright yellow flowers and small, pinnate leaves. Pinnate leaves are compound leaves with leaflets on either side of the stem. A typical feature of the paloverde (especially of the species *Cercidium microphyllum*) is its greenish bark. It contains chlorophyll, which allows the paloverde to continue the process of photosynthesis even during the dry months, when all its leaves drop off. Photosynthesis allows plants to produce carbohydrates from carbon dioxide and water by using sunlight absorbed by chlorophyll molecules. In this way, a plant turns solar energy into chemical energy.

Other common trees of the lower desert include mesquite, which can grow as high as 50 feet (15 m), and the catclaw, so called because of the curved spines on its branches. Both of these plants have pinnate leaves and yellow flowers. They produce pods full of seeds that are eagerly sought after by numerous animals and humans. The Papago Indians grind the seeds, particularly those of the mesquite, into a nutritious flour. The open forests of mesquite and catclaw often stretch out onto flat plains, beyond the rocky slopes favored by saguaros.

Another common plant of the low desert is the California fan palm, found in southern California, Baja California, and the extreme southwestern Arizona. This palm has broad, fan-shaped leaves up to 6 feet (2 m) long and almost as wide. It blooms in May and June, with sweet-smelling flowers grouped in yellowish clusters up to 10 feet (3 m) long. The fruit is edible and an important source of food for some animals. The fan palm can grow to 50 feet (15 m) tall and live for two hundred years. It is very hardy, being able to grow in the rockiest canyons and even from cracks in bare rocks.

A peculiar type of forest is found in Baja California. Here some trees have thick trunks in which they store water. Water loss due to transpiration and evaporation is reduced to a minimum. These plants are called "elephant trees"

This fantastic landscape, dominated by sparse yucca trees, also called "Joshua trees," is found in Joshua Tree National Monument in southern California. Plants belonging to the genus *Yucca,* widely distributed in the arid regions of North and South America, can be tree-shaped (like the ones in the picture) or bush-shaped. In both cases, there is a tuft of leaves on top of the plant. The leaves are narrow, with a sharp tip.

because of their thick, stubby stems. The stems give rise to long, thin, and pliable branches with very small leaves. These strange plants do not grow much taller than 13 feet (4 m), but their branches may spread to a diameter of 50 feet (15 m).

Joshua trees, which are a kind of yucca tree, are most common in the Sonoran Desert of Baja California but are also scattered throughout the deserts of the southwest United States. They can reach a height of 26 feet (8 m) with a trunk diameter of 20 inches (50 cm). Their coarse bark gives them a shaggy appearance. Joshua trees and other yuccas have formed a curious symbiotic alliance with the yucca moth. This insect pollinates the yucca flowers by gathering the pollen into a ball and depositing it, together with its own eggs, in another yucca flower. After pollination has occurred

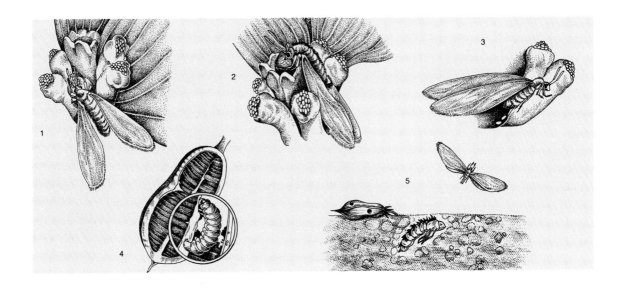

An interesting symbiotic relationship involves the yucca and the yucca moth. While laying its eggs, the moth helps pollinate the plant, which, in turn, provides food for the the growing larva. Starting at the *top left,* the female moth takes pollen from a yucca flower (1), shaping the pollen into a pellet that it will carry to another flower (2). Here it lays one or two eggs (3). The eggs hatch, and the larva feeds on the seeds of the plant (4). When the larva has completed its growth, the host pod falls to the ground. The larva goes into its pupa stage and remains underground. When it has completed its metamorphosis into a moth, it will dig itself out, helped by the hooks on its back (5).

Following pages: An incredible burst of colors is typical of the vegetation of the area bordering Death Valley. When traveling through the North American deserts during the spring rainy season, the sudden blooming of these areas, usually arid and dull, can be a beautiful surprise.

and seeds have formed, the eggs hatch and the larvae feed on the same yucca seeds which their parents helped create. But enough seeds survive so that new plants will be able to grow.

Probably the strangest looking plant in Baja California is the cirio. Standing 82 feet (25 m) tall, it resembles a gigantic upside down carrot. Its tall trunk may bend into gracefully arched shapes or even into loops. Water is stored in the thick, lower part of the trunk which has no branches. When it rains, though, tufts of small, waxy leaves project from the trunk. The waxy outer skin of these leaves is a perfect water-saving device. In dry periods, these leaves are shed. Yellow flowers appear at the top of the cirio's trunk, clustered in a rosette. So curious is the cirio that it has been called the "boojum." This term for a strange being from a distant land comes from Lewis Carroll's famous nonsense-poem, *The Hunting of the Snark.*

A smaller relative of the cirio is the ocotillo, found in the lower Sonoran and Chihuahuan deserts. At first glance, this plant appears to be a tangle of dead twigs stuck in the ground. Actually, it is one of the plants best suited to desert life. A full crop of small leaves will develop in just forty-eight hours after a rain. Unlike other desert plants, the ocotillo has fleshy, tender leaves, with no special devices for coping with arid climates. The leaves drop off as soon as water is no longer available. But the ocotillo can develop leaves at any time of the year, taking advantage of any and all rain. In a single year, it can produce leaves three or four

times. Moreover, the ocotillo, like the paloverde, has chlorophyll in its stems so photosynthesis can continue, although at a reduced rate, throughout the year.

Annual flowering plants, unlike all the species described so far, manage to survive dry periods by means of extremely resistant seeds. As soon as rain falls, the seeds germinate. The new plants grow rapidly, flower, produce seeds, and die, all within a few weeks. Hundreds of species of this type exist in all deserts. In spring, if winter rains have provided enough water, many parts of the desert burst into a kaleidoscopic show of flowers of all possible colors.

Some desert plants adopt yet another strategy—they endure dry periods. A typical example of this is the resurrection plant, found in southwestern Texas and in the Chihuahuan Desert. This remarkable species manages to survive even in the hottest desert areas, where very few other plants can live. It can tolerate almost complete dehydration. It can be reduced to 3 percent of its normal amount of water and when in this condition, the plant appears completely dead. When the rains come, though, it will revive, or "resurrect," turning green within a few hours. This is possible because of the plant's efficient water-absorption mechanism that includes a system of shallow roots as well as stems and leaves that can take up water directly into the plant. During dry periods, leaves and stems are brown and tightly curled, but as the plant absorbs water, they stretch out due to the expansion of the cells on the upper surface of the stems. The uncoiled leaves expose their chlorophyll-rich surfaces to sunlight, and the first sign of photosynthesis can be detected only three hours after rain. The plant continues its metabolic activity for two to five days until all available water evaporates. Then the plant dehydrates again and becomes dormant. It can remain in this condition for as long as two or three years.

Yet another notable plant of the American deserts is the creosote bush. It is found throughout the Southwest, except in the higher and colder areas. It has small leaves which drop off during dry periods. It can grow to 5 feet (1.5 m) tall. Its peculiar feature is its means of reproduction. It can form clones, or identical copies, of itself that expand outward from the center. Often the initial, central bush will die and leave a circular colony of new bushes. Some colonies of creosote bushes have been estimated to be five thousand years old.

SMALL ANIMALS OF THE LOW DESERTS

The fauna of the low deserts are no less rich than the flora. During the day, the air is filled with the buzzing and chirping of myriads of insects. In spring, when wildflowers are blooming, huge numbers of butterflies can be encountered. There are very many different species of butterflies, from the tiny marine blue to the slightly larger "snout butterfly," which has long mouth parts that resemble a bird's beak. In canyons, along streams, and next to roads, the beautiful sulphur with its bright yellow wings can be found. One larger species is the monarch butterfly, with black and orange wings. The monarch's colors warn possible predators: this butterfly is not safe to eat. This is because the monarch accumulates irritating substances in its body when it is a caterpillar.

The Chemical Defenses of Insects

The monarch butterfly is not the only insect that defends itself with chemicals. Blister beetles, when disturbed, secrete a toxic substance called "cantharidin." This chemical produces blisters on human skin, and it certainly causes distress to any animal that tries to eat such beetles. The adult beetles feed on flowers and nectar, but the larvae prey on grasshopper eggs. Other species lay their eggs under flower buds. When the flowers bloom, the newly-born larvae attach themselves to bees when they enter the flowers in search of nectar. In this way, the larvae hitch a ride to the hive, where they eat bee eggs and pollen.

Another typical desert insect that employs a chemical defense against predators is the Pinacate beetle. This is a small, shiny black insect, no longer than 0.8 inches (2 cm). Pinacate beetles roam all over the desert. If they are disturbed, they stand on their heads and secrete a foul-smelling substance from their abdomens. This is usually enough to discourage any predators from eating them. This trick does not always work, however. In some areas, deer mice have learned to defeat the beetles and get themselves a tasty meal. A mouse approaches the beetle head on, grabs it, and swiftly thrusts its abdomen into the sand, thus avoiding the disgusting spray. The mouse can then eat the front part of the beetle at its leisure.

Eight-Legged Poisoners and Biters

Since insects are so abundant in the desert, it follows that invertebrate predators are also present. Spiders and myriapods (centipedes and millipedes) are common.

Opposite: A colorful group of monarch butterflies lights on a branch. Besides being able to accumulate toxic substances as a defense, this butterfly also makes incredible migrations. During the winter, it flies for thousands of miles from the mild regions of the United States all the way to northern Mexico. Here, immense groups of butterflies concentrate in a few forests in the Sierras.

The drawing shows the main stages in the capture of a tarantula by a hawk wasp. *Starting at the top left:* When ready to lay its eggs, a wasp spots a tarantula suitable for its needs and starts to dig a hole nearby. The hole is about 10 inches (25 cm) deep. From time to time, the wasp stops working to make sure that the tarantula is still around. When the hole is ready, the wasp stings the tarantula, injecting a paralyzing substance into its body. Then it lays one egg in the spider's abdomen and drags the animal inside the hole. Next, it closes the hole. When the egg hatches, the larva will grow, devouring the spider, which is still alive but unable to move.

The large, hairy tarantula is the most striking of these creatures. The female's body grows to 2 inches (5 cm) long, and the entire animal can measure up to 6 inches (15 cm) from leg tip to leg tip. Despite its fearsome appearance, this huge spider is actually a shy and quiet creature. It spends the day hidden in its hole and comes out only at night to hunt grasshoppers and other large insects. It will certainly not bite unless it is picked up with bare hands or severely harassed. In any case, its bite is painful but not very harmful to people.

Even the huge tarantula is not safe from predators. Its principal enemy is a large blue and red wasp, called the "tarantula hawk." It can be up to 1.5 inches (4 cm) long. The female wasp attacks, stings, and paralyzes the tarantula with a venom. The wasp drags the spider into her burrow and lays a single egg on its paralyzed body. When the egg hatches, the wasp larva will feed on the spider, grow to adulthood, and emerge from the ground the next spring. By stinging and paralyzing the spider, the wasp provides fresh food for its larva. Although the idea might sound gruesome, the spider will literally be eaten alive. The males of this species do not hunt tarantulas and do not even have a

Some poisonous animals have peculiar sexual behavior patterns, intended to avoid direct contact during mating. This contact could be hazardous to their survival. The male scorpion, for example, deposits a spermatophore (a packet containing sperm) on the ground. The female then picks it up during the mating dance. During the dance, the two scorpions grab each other by the pinchers. Some experts think that this strange behavior is the result of two opposing needs for the male. On one hand it has to get close to the female during mating, but on the other hand it must protect itself against an attack by the female.

stinger. They feed on nectar and, after coming out of the ground in March or April, mark a territory which they defend and to which they attract females for mating. In May or June, after mating, the males die.

In addition to tarantulas, two other species of desert spiders are particularly worthy of notice. The first one is the black widow, which is common throughout the arid regions and is found in quiet recesses, brush piles, and sometimes also in houses. It is usually timid and will bite only when trapped or threatened. Of the two sexes, only the female is poisonous. Her bite is not especially painful, but severe muscle and abdominal pain may follow, as well as, strangely enough, soreness on the soles of the feet. Victims may sweat profusely and have swollen eyelids. In most cases, though, they recover after a few days of severe suffering.

The second species of spider is called the "brown recluse." It is no longer than 0.4 inches (1 cm). It lives in a similar habitat to the black widow but is more likely to be found inside houses. This spider, too, is not aggressive and will bite only to defend itself. The wound from its bite may not heal for several months. Otherwise, the victim suffers no serious symptoms.

Desert scorpions, on the other hand, can inflict dangerous stings, much more so than spiders. Many species of scorpions live in the American deserts. Some are more dangerous than others. The sculptured scorpion, one of the most poisonous, reaches 2 to 3 inches (7 to 8 cm) in length. The striped-tail scorpion is about the same size. Its sting is quite painful, but it is less dangerous than the sculptured scorpion. As a rule, scorpion stings are extremely painful, but in most cases they are not deadly.

Huge millipedes, nearly 5 inches (12 cm) long, and giant centipedes, up to 6 inches (15 cm) long, also inhabit the desert. Centipedes have twenty to twenty-three pairs of legs—not one hundred as their name suggests—and run across the desert floor like miniature trains. During the daytime, they hide in holes in the ground. When night falls, they come out to hunt insects and other invertebrates. Sometimes, large individuals even catch and eat small toads and lizards. The female takes great care of its offspring, curling up on top of its cluster of eggs and frequently moving them with its mouth, perhaps to clean them. This animal must be treated with care, since its bite is quite dangerous.

Some desert insects seek warm-blooded creatures, in-

A black widow, seen here in its web, is considered one of the most dangerous spiders in the temperate-hot regions of the Western Hemisphere. In most cases (95 percent), though, its bite is not lethal to humans.

cluding humans. The chigger is particularly irritating. Adult chiggers, just 0.1 inch (3 mm) long, hunt insects. The smaller larvae, however, attach themselves to passing animals and suck their blood. In humans, chigger bites can cause an irritating itch for several days or weeks.

Reptiles

In western North America, there are about 140 species of reptiles. Most of these live in the deserts of the Southwest. Many of these creatures, particularly the lizards, are diurnal, or active during the day, for most of the year.

The large chuckwalla, a member of the iguana family, can grow to be nearly 16 inches (40 cm) long. It is dark gray, with a fat belly and thick skin that hangs in folds. It is mostly vegetarian and prefers rocky hillsides. When alarmed, it will scurry into the nearest crevice and inflate itself by gulping air. In this way, the chuckwalla gets tightly wedged into the crack, and it becomes almost impossible to get it out without seriously injuring the stubborn lizard.

Other typical lizards include the zebra-tailed and the

Some typical small animals of the lower desert are: *(from left to right, back row)* a chuckwalla, a small desert fox, a zebra-tailed lizard, a fringe-toed lizard *(front row)* a horned lizard, a leopard lizard, and a desert tortoise.

fringe-toed lizards. The former, about 8 inches (20 cm) long, is one of the world's fastest reptiles. It can run at 15 miles (24 km) per hour. It feeds mainly on insects and has a habit of running and stopping briefly to whip its showy striped tail from side to side. The fringe-toed lizard can grow to 12 inches (30 cm) long and is particularly adapted to living in loose sand. Its long toes are equipped with delicate fringes that provide traction on the dunes and while hunting for insects. If it senses danger, this lizard can bury itself in the sand and "swim" for short distances just below the surface.

Another notable member of the iguana family is the leopard lizard. Reaching a length of 12 inches (30 cm), it has a handsome pattern of dark spots on a sandy background. Like the leopard after which it is named, this lizard is a swift predator. It can hunt and kill many kinds of small animals, including other lizards.

Among the desert lizards, the horned lizard, or horned toad, deserves mention. It owes its name to the tough crest of horns and plates around its neck, and to a series of short spiny scales on its back and tail. The horned lizard is a

A Gila monster rests in a vigilant posture. The two species of the genus *Heloderma* are the only venomous lizards living today. They are dangerous only if disturbed. Otherwise they are shy, quiet creatures that are hard to find in the tangle of rocks and bushes where they live.

solitary animal. It is often found on rocky hillsides close to nests of ants, one of the horned lizard's favorite foods. A squat shape and camouflage coloration make these animals hard to spot on rocky ground. Species in the lower desert are about 5 inches (13 cm) long. Females can produce up to twenty little lizards in a litter which they bear live.

The most extraordinary of the desert lizards is certainly the Gila monster. It can grow to 2 feet (60 cm) in length. Its beaded skin has blotches of pink, yellow, or orange on a dark gray background. This bright coloration is actually a warning signal. In fact, the Gila monster is one of only two species of poisonous lizards living on earth today. Gila monsters, however, do not inject their poison through fangs, as snakes do. Their poison is produced in glands in the lower jaw and is released along grooves in the teeth. After biting their victims, Gila monsters do not let go, but hang on and chew the venom into the victims' tissue.

Actually, Gila monsters use their poison primarily to defend themselves, not to kill prey. They are rather sturdy animals and can easily obtain food such as small mammals,

Rattlesnakes, even though they are among the most dangerous snakes on earth, are not especially aggressive toward each other. Seldom will two males fight using their poisonous fangs. Duels, when they occur, look more like a dance. The drawing illustrates the main stages of one such duel. This is a ritualized duel, in which the two snakes cling together, swaying the front part of their bodies and facing each other. In the end, the winner lies on top of its rival.

reptiles, and eggs without having to use their venom. They are slow creatures and move in a waddlelike fashion. They are shy and hunt at night, so they are not often seen. To find them, you have to look into their daytime shelters, which are cracks in rocks, caves, or burrows. Being so elusive, they are not very dangerous to people. If molested, though, they react with amazing swiftness and latch onto a hand or an arm. They are particularly fond of water and after rains may bathe in a pool for many hours. They are not active in winter but live on fat stored in their thick tails.

Snakes

Among the many snakes living in the desert, the most familiar are the rattlesnakes. The largest species is the diamondback rattler. Most individuals are about 4 feet (1.2 m) long, but some reach a length of 6 feet (2 m). Rattlesnakes belong to the family of pit vipers, so called because of two depressions between their eyes and nostrils. These pits are extremely sensitive to heat and are used to detect warm-blooded prey during night hunting. They are so efficient that a rattler can actually detect and seize its victim even in total darkness.

A western diamondback rattlesnake is seen up close. In the picture, it is easy to see the "pits" under the nostrils. The pits are sense organs that enable the snake to detect small warm-blooded animals from a distance. This snake preys mainly on mice, other rodents, and young birds in nests on the ground.

All rattlers are poisonous. They inject venom through a pair of long, hollow poison fangs on the upper jaw. When their mouths are shut, their fangs are folded backward and rest on the roof of the mouth. But when they attack, the poison fangs are quickly lifted into an upright position. An attacking rattler will dart forward as quick as lightning and then retreat to its coiled position. It can strike several times in just a few seconds, especially when frightened or when trying to kill large and active prey.

When threatened or alarmed, rattlers will shake the ends of their tails, making the sound for which they are famous. At birth, the rattle is only a simple button. Every time the snake sheds its skin, a new segment is added to the rattle. The string of rattles can grow to be several inches long. But when it gets over a certain length, it may start to break off. The rattling, which sounds like a loud buzz, is a warning signal. If the intruder moves quietly out of the way, the snake will not follow. In fact, rattlesnakes are not very aggressive and would rather avoid danger than attack. When cornered or unintentionally disturbed, though, they can be dangerous.

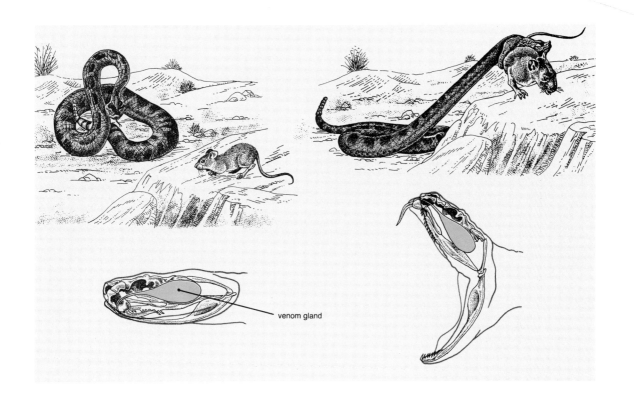

When it comes across prey *(top left)*, the rattlesnake stays still and, lifting its head slightly, darts its tongue in and out. The prey, a kangaroo rat in the drawing, tries to escape, but the rattlesnake attacks the rat *(top right)* and devours it. These snakes are equipped with a pair of fangs which, with their mouths shut, are bent backwards and rest on the roof of the mouth *(bottom left)*. When the snakes open their mouths, the fangs rotate outward *(bottom right)*, preparing to seize the victim and inject the poison. The venom is stored in a poison gland, which is connected to the fangs through a small channel.

Among the rattlers, there is a small species, 2 feet (60 cm) long, which is perfectly adapted to living in sandy deserts. It is the so-called sidewinder rattlesnake. It owes its name to its peculiar winding movements along the sand. It moves sideways so that, at any moment, only two points of its body are touching the ground. It leaves a characteristic track on the sand, like a series of parallel J shapes. Another identifying feature of the sidewinder is the pair of "horns" above its eyes. The sidewinder, like all other rattlers, spends the winter hibernating and gives birth to its young in summer.

The Desert Tortoise

Turtles are not common in the desert, except where there is a permanent source of water. One species, though, has adapted to a totally terrestrial life. It is the desert tortoise, found on hot rocky slopes or even on flats among cacti. It can be as long as 14 inches (35 cm) and has a rough, deeply-grooved shell. Its legs, neck, and head are thick and sturdy, and its mouth is equipped with a horny beak. It is almost exclusively vegetarian and derives all the water it needs from the food it eats.

BIRDS OF THE LOW DESERT

Birds are among the most visible animals living on earth, and even in the desert they are in evidence everywhere. At dawn, the coolest time of the day during the hot summer, they burst into activity. Doves start cooing, finches sing, and other birds of varying shapes and sizes add their voices to the morning chorus. Around midday, when the temperature rises well above 100°F (38°C), all activity stops. It starts again, as intense as before, near sundown.

Wrens and Thrashers

A common and conspicuous desert bird is the cactus wren. It is easily recognized by its large size, as the bird can grow to 6 inches (16 cm). It has broad white eye stripes, a densely speckled breast, and a long banded tail with a white tip. Unlike its smaller relatives, the cactus wren does not always carry its tail cocked upward. Its song is a series of not very musical rattles that are among the most familiar sounds in the desert in summer. It builds its nest, which is shaped like a dome, in the branches of a cactus, such as a cholla. The cactus itself acts as a defense against predators.

Among the most common ground birds in the desert is the curve-billed thrasher. Its characteristic call can be heard all over: whit-whit-wheet. It also has a musical song consisting of a long series of notes. The curve-billed thrasher is a rather large, brown bird with a long, slightly down-curved bill, and bright orange eyes. It feeds mainly on insects and other invertebrates found on the ground. It builds a nest, shaped like an open cup, in the thick tangles of mesquite or cholla branches. It lays its eggs in spring or early summer during the rainy season. The eggs are light blue and softly speckled.

Doves, Quails, and Roadrunners

The white-winged dove flies north from Mexico in summer and can be abundant in the Sonoran Desert. It is 10 inches (25 cm) long, sandy colored, and displays white patches on its wings and tail when in flight. The dove's cooing is another familiar desert sound. Its nest, like those of most pigeons and doves, is a flimsy platform made of twigs and built in a tree or a shrub. The female lays two white eggs in spring and summer. The white-winged dove is one of the most effective agents of propagation for the saguaros cactus. The doves drink the water that accumulates in the saguaro's yellow-white large flowers, thus helping in pollination. They also eat large quantities of saguaro fruit and release

Opposite: The Gila woodpecker is one of the most unexpected desert inhabitants. It belongs to a group of birds that feeds on insect larvae which they find under bark or in rotting trunks. This particular species has learned how to take advantage of the large cacti growing in the American deserts. These cacti are as large as trees, and they provide plenty of food.

the tiny seeds with their droppings, often very far from the mother plant. The seeds pass undamaged through the dove's digestive system.

The wide expanses of the desert are inhabited by some true ground birds. Among them, a particularly striking example is the roadrunner. This large bird belongs to the cuckoo family, and with its long tail, it can reach 22 inches (55 cm) in length. It has a dark-striped coat of feathers and a shaggy crest on its head. The roadrunner has strong legs. Even though it can fly, it prefers to run after its prey, which includes lizards, snakes, and large insects.

The song of the roadrunner resembles that of a dove and consists of coos that descend in pitch. It also makes a peculiar rattling sound by rubbing the two parts of its beak together. This bird also builds its nest in a cactus or a dense bush. The nest is shaped like a shallow cup and made with twigs. In it, the roadrunner lays three to eight white eggs.

The roadrunner is superbly suited for living in the desert and can survive on very little water. This bird has a bare patch of dark skin on its back. On cool mornings, especially in winter, the bird exposes this area to the sun's rays, perhaps to absorb heat.

Another common ground-dwelling bird is the Gambel's quail. This bird lives in groups of twenty or more individuals that continually search the desert floor for seeds and insects. It is 10 inches (25 cm) long and quite handsome. The male has a gray back, a black throat patch bordered by white, a reddish crown, and a buff-colored abdomen with a distinct black spot. The most distinctive feature, though, is one black, curved plume, shaped like a teardrop, on top of the bird's head. The plume bobs up and down as the bird walks. Females resemble males, but their markings are less distinct and their head plumes are a little smaller.

The territorial call of the male Gambel's quail is a unique crowing that is repeated many times, especially in the first hours of the day. During the mating season, groups of quails break up into breeding pairs. The female makes its nest on the ground. Ten to sixteen speckled, well-camouflaged eggs are laid in it. The young birds are ready to leave the nest and go in search of food soon after they are born. Within two weeks, they can fly.

Dwellers in Cactus

Woodpeckers abound in North American deserts, especially in the mesquite and saguaro forests. A common

A mockingbird feeds on the sugary pulp of a prickly pear fruit. Many desert plants bear nutritious fruit, allowing a large number of animals to survive in arid environments even when they are not particularly suited for it.

Opposite: The roadrunner is one of the fastest ground runners. With its head stretched forward and its tail in a horizontal position *(top drawing),* the roadrunner can reach a speed of 15 miles (24 km) per hour. While running, it balances and can change direction by using its long tail like a rudder *(two middle drawings).* It can also stop suddenly, fanning out its tail and lifting it up *(bottom drawing).* The roadrunner's toes are in an X pattern so the bird can grip the ground firmly while running.

species is the Gila woodpecker. Its breast and belly are sand-colored, and its back is striped black and white. The male, as in many other woodpecker species, has a red crown, a feature absent in the female. When in flight, both sexes show large white markings on their wings. The call of these birds is a monotonous and repeated pit-pit-pit, yet another familiar desert sound. Gila woodpeckers, together with other desert woodpeckers such as gilded flickers, dig holes in trees and in saguaro cacti. Each couple may dig many such holes in a year, even though they lay their eggs in only one of them. Many other animals end up using the extra holes. One example is the elf owl, a sparrow-sized bird with distinctive white eyebrows and fierce yellow eyes. The call of this miniature bird of prey is a rapid series of notes, somewhat like the yelps of a puppy. Elf owls eat insects and are strictly nocturnal animals. They lay three or four eggs in nests inside holes in trees and cacti.

Other much larger nocturnal birds of prey dwell in the desert. The largest is the great horned owl, which can be as long as 20 inches (50 cm) and have a wingspan of 5 feet (1.5 m).

Its deep and gloomy hoots are often heard at night. It preys on rabbits, skunks, and perhaps on some birds and reptiles. These owls build their bulky nests in trees or large cacti, often in full view.

Small Perching Birds

Small perching birds are also well represented in the desert. Among them is another classic species of arid environments, the black-throated sparrow. This is a small bird, 4 to 5 inches (10 to 13 cm) long. Its back is gray, and its face has a bold black-and-white pattern. This bird can survive with little water and actually satisfies most of its requirement for water from the seeds, plants, and insects it eats. This bird can be found all over the desert, even in the sun-baked flats, among the creosote bushes and sagebrush. Its call is a sequence of sweet notes and trills and can be heard throughout the region, especially after rain. Its nest, shaped like a cup, is built in a dense bush or in a cactus. Usually the bird lays three or four white eggs.

Another desert songster is the house finch. This bird is abundant around towns and dwellings as well as in the most remote desert areas. Males have bright red heads and upper chests and, in flight, they often display reddish rump patches. Females, on the other hand, are a grayish brown color. The long, warbling song of these little birds can be heard throughout the year, although more frequently in spring and summer during the breeding season. Their nests are also shaped like a cup and may be built on cliffs or buildings, and also in trees or cacti. Four or five spotted, blue-green eggs are laid. Like many other desert birds, the house finch must have some water other than what it gets with its food. Fortunately for birds, they can fly some distance to water holes during the dry season.

A peculiar bird is the phainopepla, the only North American species of the silky flycatcher family. The male, about 7 inches (18 cm) long, is shiny black, with a long tail and a distinct crest on its head. It has deep red eyes. In flight, it shows large white markings at the tips of its wings. The females have a similar shape and pattern, but they are gray, not black. The phainopepla feeds mainly on fruit and insects. Its stomach is highly modified for a diet of fruit. The pulp is digested in about an hour, and the seeds are regurgitated. Fruit is rich in carbohydrates but it lacks proteins. In order to achieve a sufficient diet, these birds have to eat the equivalent of their body weight each day.

Opposite: Some common birds are pictured in the lower deserts of America. *From left to right:* (in flight) a gilded flicker and a male flycatcher; (perching) a great horned owl grasping a small rodent, a male black-throated sparrow, a white-winged dove, an elf owl on a blossoming cactus; and (on the ground) a family of Gambel's quails.

MAMMALS OF THE LOW DESERT

Mammals are hard to spot because they are active mostly at night. The desert, nevertheless, hosts a rich variety of mammal species, and some are specially adapted to life in arid regions. One such species is the banner-tailed kangaroo rat, which is common in the Sonoran and Chihuahuan deserts. As its name implies, it has strong hind legs and a long tail tipped with white fur that it can wave like a flag.

Life of the Kangaroo Rat

Kangaroo rats are nocturnal animals that live in burrows. Like kangaroos, they use their hind legs for jumping and their tails to maintain balance. Thus, their front legs are always free to dig and handle food. They can also use their hind legs to drum on the ground, perhaps a territorial signal or a warning when predators are near.

These rodents can go a lifetime without a drink of water. Water is as necessary to them as to any other living creature, but they get all the water they need from their food, which consists mainly of seeds and other vegetable matter. Their kidneys are incredibly efficient and produce an extremely concentrated, almost solid urine. Moreover, they also recover the moisture from the air they breathe, thanks to their highly specialized nasal passages that condense water from exhaled breath. During the day, kangaroo rats close the entrances to their underground chambers with a little dirt in order to avoid the hot midday sun. Their tunnel systems can be quite complex, forming a maze 14 feet (4.5 m) in diameter beneath the ground. One such burrow can be the home to several individuals, and it might be used by many generations of rats over several decades. The tunnels may also serve as storage rooms for seeds and other kinds of food to be eaten during the hottest part of the day or during the winter months.

Kangaroo rats reproduce in spring, and the offspring can remain with the mother in the tunnel system for some time after weaning. The males, on the other hand, tend to be rather solitary. In general, these rodents are markedly less social than ground squirrels or prairie dogs.

Other Desert Rodents

Another common desert rodent, frequently found on rocky slopes, is the antelope ground squirrel. It is 9 inches (22.5 cm) long, with a short tail and a stripe running along its sides. These ground squirrels live in holes among the rocks. Their burrows keep them safe from most predators, except

Opposite: An antelope ground squirrel, standing on its hind legs, surveys the surrounding territory. The American deserts, as well as those on other continents, provide a host of ecological niches (habitats where various species can live). These are ideal homes for small mammals, provided they adapt to desert food and develop special devices to avoid overheating and dehydration.

snakes. When a snake is near, the antelope ground squirrel will often scurry into thick tangles of cacti rather than hide in its hole.

The antelope ground squirrel feeds mainly on cactus fruit and hibernates only for a few days in winter, since the climate stays mild at this time of the year. During the torrid summer days when ground temperature may rise above 120°F (49°C), it may become completely inactive. This kind of hiberation, called "estivation," occurs during the hottest months rather than the coldest. It is a method to save water and to avoid the extreme heat.

Another common rodent of the lower desert is the round-tail ground squirrel. It has colonized the hottest regions, even the low desert flats covered with creosote bush. It measures about 9 inches (22 cm) long and has light brown fur and a short, rounded tail. This squirrel feeds mainly on green leaves and stems, and not so much on cactus fruit. As a result, it stays inactive longer than other species during dry periods.

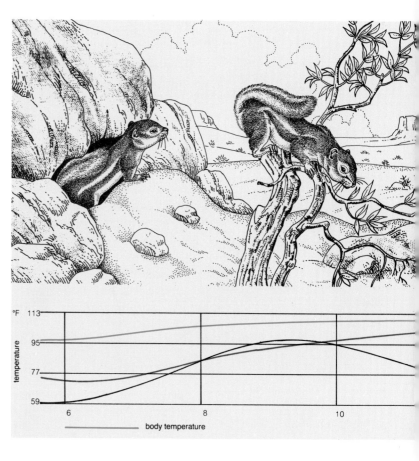

The antelope ground squirrel is one of the few animals in the American desert that can remain active during the hottest hours of the day. It actually keeps its body temperature higher than the temperature of the environment. The drawing illustrates the main stages in a typical day for this squirrel. The graph relates activities to body temperature and changes in environmental temperature.

The pack rat is probably the best-known rodent of the North American deserts. It is also called a "woodrat." This small creature, which is just 10 inches (25 cm) long including its tail, builds huge nests with any type of material it can find. It uses twigs, rocks, paper, and leaves to build nests that can be over 6 feet (2 m) high and 6 feet (2 m) wide. Sometimes the nests are built among cacti, or they can be placed in more sheltered sites, such as inside a small cave. Often, newcomer rats will remodel and reuse old dens. Very large dens, like some found in the Grand Canyon, have probably been inhabited by families of pack rats over ten thousand years.

Pack rats use their urine to bind their nests together. In the arid climate, this particular type of nest can last for centuries. The dens are a refuge against predators and, since they are not enclosed, provide a kind of "air conditioning." Other species might share the nest, including the black widow and brown recluse spiders, scorpions, lizards, and snakes.

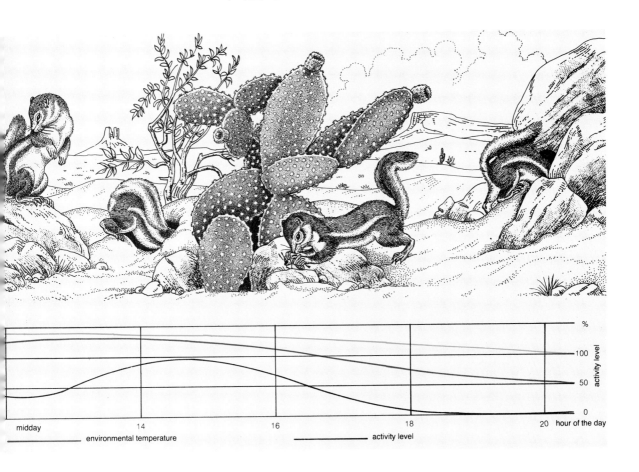

Although their shy and mainly nocturnal habits make them hard to find, many small mammals live in the lower deserts of North America, despite harsh environmental conditions. The most common mammals are shown in the drawing. *At the top,* a banner-tailed kangaroo rat bounces away with sudden leaps. *In the upper left,* a white-throated woodrat stands in front of its mound-shaped den. A desert woodrat, a so-called pack rat, is shown in front of a prickly pear cactus *on the right.* Below it is a desert shrew. Another kangaroo rat rests in its cool burrow, where it retreats during the hottest times of the day.

Pack rats feed mainly on the fleshy pods of prickly pear cactus, but they also eat mesquite seeds, fruit, and bark. Oddly enough, these animals do not even try to avoid cactus spines, and their skin is often pierced by the sharp spines. This would cause real problems for any other animal, but not for the pack rat; it simply pulls the spines out, one by one.

The Desert Shrew

One of the most surprising inhabitants of the lower desert is the tiny desert shrew, which is no longer than 4 inches (10 cm) and weighs a mere 1/6 of an ounce (5 grams). Normally, shrews live in rather moist environments. They need to eat constantly to get enough energy, since their metabolism is very high. Most of their food is animal matter, thus rich in protein. They need to drink large amounts of water to remove chemical waste left over from the breakdown of protein.

How, then, does the desert shrew manage to survive in such arid lands? The secret is simple: this shrew has learned how to create a habitat for itself in which the temperature and humidity do not reach extreme limits. Its nest, which is

built in thick vegetation, provides shelter and is also a perfect resting place for short periods in between one hunt and the next.

Moreover, this animal has quite large ears for its small size. They help lower the shrew's body temperature by releasing heat. The desert shrew feeds mainly on insects, which provide some of the water it needs. It also stores food so that it does not have to hunt during the hottest hours of the day.

Besides these behavioral adaptations, desert shrews also have highly efficient kidneys that produce more concentrated urine than that of other shrews. They can also condense and recover moisture from the air that they exhale, as a result of special nasal passages. Finally, they enter a state of sluggishness or semi-hibernation when food is scarce. All of these features make it possible for shrews to survive in the desert.

The shrew's activity reaches a peak during the mating season. The three to six offspring are born in summer and grow rapidly. After two weeks, the total weight of the litter may be twice that of the mother. Most of this increase in weight comes from nursing on milk from the mother, who has not had a drink of water. After weaning, the mother, like some other carnivores, brings food to the nest and regurgitates it in the babies' mouths.

The Peccary

One of the larger mammals of the lower desert is the collared peccary. Also known as the *javelina* (a Spanish word meaning "spear") because of its long canine teeth, this animal resembles a small pig. It weighs up to 50 pounds (23 kg) and is 20 inches (50 cm) high at the shoulder.

Actually, the peccaries branched off from true pigs around forty million years ago. The two groups are similar in the shape of their bodies and in their habit of rooting in the dirt with their snouts. Unlike pigs, however, the peccary has only a stub for a tail. Its coat consists completely of bristles, while pigs also have hairs. Peccaries have a multichambered stomach, specially suited for helping to digest material, while pigs have a simple saclike stomach. Also, the canine teeth of peccaries do not curve outward like those of pigs.

Peccaries live in stable social groups of up to twenty individuals, with a similar number of males and females. Females give birth to only two piglets, which are soon

Following pages: A collared peccary emerges from the bush. Although peccaries resemble small boars, they are not closely related to the true pigs. The peccaries and pigs began different evolutionary branches about forty million years ago, around the end of the Eocene epoch.

independent. This is another difference between peccaries and pigs. Pigs have more numerous offspring that require parental care for much longer. Males and females look very much the same. Members of a group keep in touch with each other by making grunting sounds. Their territory covers about 1 sq. mile (3 sq. km). They mark the boundaries of their territory with a secretion from scent glands located on the back near the tail. Members of a group also rub each other with this secretion.

In winter, peccaries look for food mainly during the daytime. At night they huddle close together to keep warm. In summer, on the other hand, they are active mainly at night. During the day, they rest in the shade. Their eyesight is poor, but their hearing and sense of smell are quite good. They feed on any type of plant matter, including the prickly pear cactus, which they eat spines and all.

Young peccaries do not leave to join other groups, so members of a group may be closely related. As a result, females often nurse youngsters other than their own, and males do not seem to compete for females. The social behavior of these animals raises important questions for biologists studying evolution. For example, how do these animals manage to avoid inbreeding between close relatives? More research is needed before such questions can be answered.

The Kit Fox

Carnivores are widespread in the low deserts, although most mammalian predators are found at higher elevations. An exception is the kit fox, a small member of the dog family. It is usually no longer than 30 inches (75 cm), weighs about 5 pounds (2.25 kg), and has huge, batlike ears.

Kit foxes live in pairs in dens dug in the soft soil. They feed on ground squirrels, mice, and insects. As is often the case in the dog family, the male has an important role in raising the offspring. The speed and agility of the male make him a fine hunter, capable of catching ample food for his mate and their pups.

THE HIGH DESERT

From any spot in the low deserts of North America, mountain ridges and plateaus rising above the desert plain can be seen. At higher elevations, the vegetation changes dramatically. The forests of giant cacti and the mesquite woods give way to wide stretches of grasslands, to slopes covered with junipers and oaks, and to canyons. To the north, past the Mogollon Rim, desert areas turn into rocky expanses, as in the Painted Desert and Monument Valley. Here the vegetation is sparse, but elsewhere, to the west and further north, wide stretches of sagebrush are found.

The Pronghorn

A century ago, most of the arid Southwest was covered with rolling grasslands and pockets of woods near water. Recently, overgrazing by domestic livestock has damaged what was once a rich habitat. Many species that once thrived here have seen their numbers decrease or have disappeared altogether. Today, large areas of desert grassland exist only in southwestern Texas and northern Mexico. Such prairies were once home to large herds of the southern pronghorn. Although this species shares similarities with antelope and goats, it is included in a family of its own. Today, southern pronghorns are very rare in southern Arizona and New Mexico, and they are frequent only in the more remote areas of Mexico. They are highly social animals, and in the past were found in herds of several hundred animals.

Pronghorns are rather small, growing in length to about 4 feet (1.3 m). They are buff-colored with white bellies and black stripes on the head and face. When they run, they display a tuft of white hair around the tail. They live in open country, and their eyesight is amazingly sharp. Their eyes bulge from the sides of their heads to give them a wide angle of vision. At the slightest sign of danger, they take to their heels at an incredible speed of up to 40 miles (65 km) per hour. They can only maintain this speed for a short time, but it is helpful for outdistancing predators.

The mating season starts in early autumn, and males engage in fierce fights to protect their harems. The horns used by males in battle are rather short, growing only about 8 to 10 inches (20 to 25 cm), and are curved inward at the tips. Each horn has a flat prong projecting forward. Unlike the true antelope, which do not shed their horns, the pronghorn sheds the horny outer sheath of the horns but keeps the bony core. This core will be covered by a new

Opposite page: The American pronghorn is one of the oddest animals to be encountered on the wide grasslands of North America. At first sight, it might resemble an antelope. It is equipped with a kind of horn that is not found in any other family of living hoofed mammals. Its physical features are a peculiar mix and seem to belong to a diverse group of animals such as cattle, deer, and giraffes. But their odd appearance doesn't keep them from running like the wind across the grassy plains.

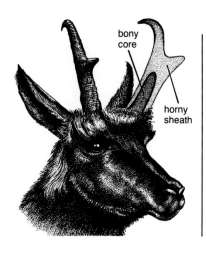

sheath before the start of the next mating season. The young are born in spring and are soon able to follow their mothers, even though, in case of danger, they prefer to hide.

Although the southern pronghorn is threatened with extinction, the northern form is holding its own. Suitable habitat for the northern pronghorn still exists in the sagebrush areas of southern Oregon and Idaho and in some parts of Nevada.

Deer

Two species of deer are still fairly common in the Southwest: the mule deer and the white-tailed deer. Neither

Opposite page: In its family, the pronghorn is the only species which is most closely related to the cattle and deer families. The males have unbranched horns with hooked tips. The horns are covered by a horny sheath, which is shed and regrown annually.

Below: The bulky structure of this buck mule deer clearly explains its common name. Mule deer, along with white-tailed deer, are among the most common hoofed mammals in North America. They can be found both in forests and in the open, provided there is enough food available.

of them is a true grassland deer. They live throughout the region, both on the desert floor and at higher elevations. They even reach the fringes of the grasslands, especially along canyons and along forests near rivers. Both species range over much of North America, but the white-tailed deer is more common in the eastern regions, while the mule deer prevails in the west.

The two species of deer are about the same size, standing just over 3 feet (1 m) at the shoulder and weighing 90 to 155 pounds (40 to 70 kg). In both species, individuals living in the desert are somewhat smaller than those of the northern and western regions.

It is easy to identify the white-tailed deer because, at the slightest sign of danger, it will show the white underside of its tail. On the other hand, the tail of a mule deer always has black on the upper surface and a solid black tip. White-tailed deer are now found mainly at high elevations in open woods, canyons, meadows, and at the edges of grasslands. Mule deer often enter the lower desert, but their distribution overlaps with that of the white-tailed deer at middle elevations.

These deer feed on twigs, leaves, and occasionally even on cacti. They also like grass and the fruit of mesquite. During summer, they are active only at dusk and at dawn, and spend the hottest hours of the day in the shade. They are known to dig with their front hooves in the pebbly beds of dry streams for water beneath the surface.

In the fall, during mating season, the males of both species establish harems of does. Fights for possession of does are often seen at this time. In December or January, the males shed their horns. In spring, the females give birth to one fawn with camouflaged fur. With their reddish brown coats speckled with white spots of various sizes, fawns of white-tailed deer are adorable.

Jackrabbits

The grasslands and the nearby woods are also home to many smaller grazing and browsing mammals. One of these is the black-tailed jackrabbit. This species is not just found in this region. It is frequent in most of North America, and it lives anywhere it can find the plant matter it needs to survive.

Another rabbit that is native in this region is the antelope jackrabbit. At first sight, it resembles a rabbit with long legs and absurdly long ears. It is a large rabbit, up to 20

A group of white-tailed deer marches across a snowy Utah plateau. The white-tailed deer is widely distributed in North America in a large variety of environments. On the semiarid western plateaus, these animals find ideal conditions where there is an abundance of grass. They easily adapt to feeding on twigs and cacti and can stand extreme heat or cold.

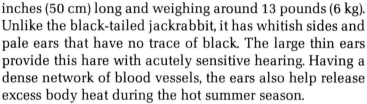

The North American antelope jackrabbit is very similar to the black-tailed jackrabbit, but it has longer ears and a black patch on the tail. On the other hand, it displays a patch of white hair on its rump. This is especially visible when the animal is moving. According to some scientists, it is meant to distract predators.

inches (50 cm) long and weighing around 13 pounds (6 kg). Unlike the black-tailed jackrabbit, it has whitish sides and pale ears that have no trace of black. The large thin ears provide this hare with acutely sensitive hearing. Having a dense network of blood vessels, the ears also help release excess body heat during the hot summer season.

A curious feature of the antelope jackrabbit is the white patch of fur that can be swiveled from one side of the rump to the other. When frightened, the hare flees diagonally, with the white spot aimed at its chaser. After about 80 feet (25 m), it suddenly switches direction and also moves the white patch so that it again is aimed at its enemy. The swift zigzag running, while always keeping the spot facing the foe, might serve to confuse a predator. The white spot, however, also marks the exact location of the animal. Thus, the function of this white spot remains a mystery.

Fortunately for the antelope jackrabbit, it is much faster than any predator in the region. Its bouncing run, with leaps covering 10 to 16 feet (3 to 5 m) resembles that of an antelope. Only when hotly chased by a predator will it break into a flat run.

Antelope jackrabbits lead a rather solitary life. They feed mainly on grass. After the rains, they eat dried grass for the rest of the year. During the dry season, they get needed moisture by eating cacti. As with other rabbits, this species has young that are born with fur and with their eyes fully open. Unlike other rabbits, though, the young (usually one to three) are hidden at separate spots in the brush, not in nests on the ground.

Carnivores

Throughout the American West, including all parts of the desert, one of the most common larger carnivores is the coyote. A member of the dog family, the coyote is smaller than a wolf. It is about 3 feet (1 m) long and can weigh 44 pounds (20 kg). The coyote is not as social as the wolf, and prefers to live alone. However, a male and female may live as a pair when there are pups to raise. The coyote also is a skilled hunter but usually seeks small prey, such as rodents, rabbits, or birds. It will feed on bodies of large animals killed by other predators and also is fond of berries and mesquite seeds. This varied diet, as well as its skill in digging for water in dry streambeds, accounts for the coyote's ability to live almost anywhere. It communicates by barking and yelping. Its eerie wailings have startled many visitors new to the

The black-tailed jackrabbit is one of the most abundant rabbits in North America. This animal, some 27 inches (70 cm) long and with especially strong hind legs, has extremely long ears in relation to its body. Interlaced with a dense network of blood vessels, they clearly act as radiators and help the animal release excess body heat.

The coyote is one of the most successful members of the dog family. Smaller than a wolf but larger than a fox, it resembles the latter in its ability to adapt to environmental changes brought about by humans. Coyotes are shot by hunters and poisoned by ranchers, yet their total population does not seem to be greatly affected by such persecution.

desert. In fact, these weird sounds are among the most familiar sounds of the American deserts.

The wolf was once one of the main predators in this region. It preyed mostly on large hoofed mammals. But as grasslands became overgrazed or were plowed for crops, the herds of hoofed mammals either disappeared altogether or sought refuge in the canyons. Wolves were forced to hunt domestic animals such as cattle. Ranchers, anxious to protect their animals, hunted the wolf relentlessly. This large carnivore is now very rare in the Southwest. In the past, it was common to the entire region, from the lower desert up to higher elevations.

Wolves can grow up to 5 feet (1.5 m) long and weigh 90 pounds (41 kg) or more. They are highly social animals, living in packs of five to ten individuals. Much larger packs were common in the past. Each group is dominated by one male and one female. These two animals are the most likely to breed and produce young. Other members of the pack bring food to the nursing female and later to the pups when they are weaned. Apart from the primates, this is one of the few examples of cooperation in raising young that is found

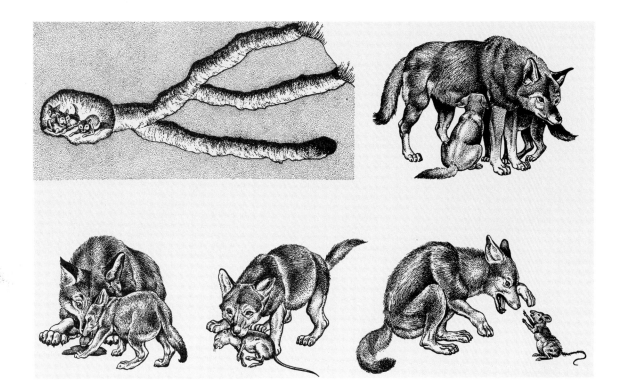

The illustrations show the phases of growth for wolf pups. *Starting at the top left:* The pups are born in an underground den with several exits. They spend the first two or three weeks of their lives here, nursing only on their mother's milk. At around the third week, the young wolves start to go out of their den to learn the many techniques they will need for their future independence. The female stops nursing and starts feeding them with regurgitated food and later with small prey that she has caught. She will feed them until they are able to hunt on their own.

in mammals. The young are born in spring, usually in caves or in secluded dens.

Wolves hunt in packs by chasing large game until the prey tires out. Then they quickly kill their victims with their powerful jaws. The prey can be a deer or, as was frequently the case in the past, an American bison. (This large hoofed mammal once lived on the eastern edge of the desert regions.) On the other extreme, wolves will often hunt small rodents and rabbits.

The home range of a wolf pack is enormous. A pack might be on the move for over eight hours every day. Communication within the pack and among different packs occurs through an amazing variety of sounds, including the famous wolf howl. They also mark their territory by urinating on rocks and trees.

Another dweller of the grasslands, as well as of the lower desert, is the badger. A member of the weasel family, the badger can be 2 feet (60 cm) long and weigh up to 24 pounds (11 kg). Its fur is grayish, with white and black stripes on the face. Its black feet are armed with large, sturdy claws that are 1 to 1.5 inches (3 to 4 cm) long. The badger is a powerful animal and a skilled digger. It lives in deep bur-

The American badger, found in the dry, open areas of the United States and Canada all the way down to central Mexico, can hold a pack of hunting dogs at bay. It feeds mainly on rodents, which it reaches by digging into their burrows. It also hunts reptiles as well as ground birds and their eggs.

rows and comes out mostly at night to hunt rodents, ground birds, eggs, insects, and reptiles. The badger is a major enemy of ground squirrels, and can easily dig up their burrows unless they are under rocks. Strong legs and dangerous claws also offer good defense, so larger predators do not attack these animals unless they are really starving. The badger leads a solitary life, joining with other badgers only during the breeding season. Two to five young are born in the spring.

Birds

Some bird species live only in the open areas of grasslands. The eastern and western species of the meadowlark meet in this region, especially in southern Arizona. Both have magnificent yellow chests with black, V-shaped marks. It is difficult to tell these two apart by their appearance, but their calls are quite different. The western meadowlark has a loud, warbling song with flutelike notes, while the eastern meadowlark's song is a slurred whistle.

Both meadowlark species feed on plants and insects. They build a nest on the ground and lay up to five white eggs speckled with brown. The western species is also common in the northern grasslands and in the sagebush deserts. They are a familiar sight along roadsides. They are easily recognized in flight by their burst of wing beats and the distinct white edges on their tails.

The white-necked raven is a large bird, rather common in arid grasslands, mesquite flats, and yucca deserts. It is usually smaller than the common raven, a bird typical of the deserts more to the north. The white-necked raven measures 16 to 18 inches (40 to 45 cm) long and is recognized by the white collar on the back of its neck. This collar, though, is not easy to see, unless the wind lifts up the black feathers that cover it, or unless the raven is intentionally displaying it.

The white-collared raven tends to be more social than the common raven. Both species prefer the same food. Actually, they eat almost anything: eggs, insects, a large variety of plants, and even carrion. Occasionally, they will attack sick or wounded animals or other birds. Their nests, made of twigs, are built on trees, yucca plants, or even on telephone poles.

In winter, huge flocks of sparrows and buntings pour onto the grasslands and surrounding areas from the northern prairies and deserts. Among the most common is the

A chipping sparrow sings while perched in a shrub. The numerous species of North American sparrows belong to a different family from the sparrows of Europe and Africa. American sparrows include the buntings and other similar birds. European sparrows are also found in North America today. They were imported from Europe over a century ago and quickly adapted to their new home.

white-crowned sparrow, with its distinctive white-and-black striped crown. Other small brown species of sparrows are difficult to tell apart solely from their plumage. Examples of these are Brewer's sparrow, chipping sparrow, and savanna sparrow. These grasslands also provide refuge for flocks of lark buntings, which flee the cold winters of the northern plains in search of seeds.

Amphibians

Summer rains fill depressions and the dry streambeds with water, which remains for a few weeks before evaporating. This is the time when the desert amphibians reproduce. It might seem impossible that any amphibians could survive in such arid climates, but actually several species have managed to do so with a good deal of success.

In the grassy plains and lower deserts of southeastern Arizona, burrowing treefrogs emerge after the first summer rains. They gather in temporary pools to breed. Their call is a hoarse squawk, repeated two or three times a second. At times other than the breeding season, these animals lead a nocturnal existence and spend the day in burrows. They

A spadefoot toad emerges from its underground hole. The adaptation to desert conditions by these strictly water-bound amphibians is truly one of the most amazing natural phenomena. It demonstrates that, through the long course of evolution, living creatures can achieve results that seem impossible.

Opposite page: The spadefoot toad manages to survive during dry periods in the desert by taking shelter in deep burrows, which it digs *(top)* with its hind legs *(bottom)*. The hind legs are equipped with special horny "spades." The toad can hibernate in its hole up to ten months a year. It resumes its activity when the rains come.

may actually remain sealed in the burrows for long periods during the driest weather. Many other desert species behave in the same way when they are far away from permanent water sources.

In the rolling mesquite grasslands of southeastern Arizona lives the Sonoran green toad. It is also found in the low creosote bush desert. This amphibian is 2 inches (5 cm) long and has striking coloration. A network of black pigment surrounds greenish yellow spots. The toad leaves its shelter only at night, when it goes in search of food. This animal also breeds during the rainy season. Then, many toads gather in temporary pools in the desert and give out their call, a panting whistle lasting up to three seconds. This

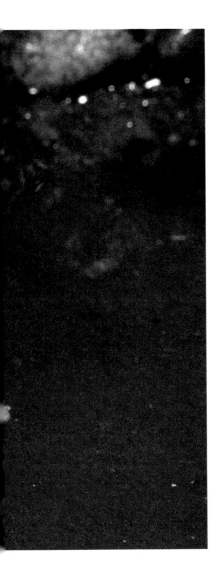

call often sounds as if it is coming from a distance, while in fact the toads may be no farther than a few feet from the listener.

Perhaps the most highly specialized desert amphibians are the spadefoot toads. Four different species are found in the desert. These toads spend most of the year buried deep in the soils of the grasslands, in the creosote bush desert, and even in salt flats and playas. They dig by using the black, sharp-edged "spade" on each hind foot. At times, though, they use holes already dug by the ground squirrels. These toads have large, bulging, yellowish eyes and a smooth brownish skin with black and light yellow spots. Most spadefoots are no longer than 2 inches (5 cm), but some can reach 4 inches (10 cm). After the first summer rains, usually within a few hours, these creatures come out from their shelters to breed in shallow pools and flooded playas. Their mating behavior is truly "explosive"—all the eggs are laid within a few days. Their calls are loud, some sounding like a low snore, others like a sheep's bleat.

The eggs hatch after two or three days, and the tadpoles turn into small toads after only two weeks. This rapid growth is quite amazing considering that for frogs and toads living in humid environments, this process usually requires several months. But water does not stay in the desert pools and on the playas longer than a few weeks. Soon the hot desert sun evaporates even the slightest trace of moisture. Spadefoot toads are thus perfectly adapted to take advantage of the short wet periods of the desert regions.

LIFE IN THE CANYONS

In the canyons and surrounding hills perhaps the richest habitats in the American deserts can be found. The canyons usually have permanent water. Consequently, rich vegetation can develop near ponds and along streams, with the varied animal life that accompanies it.

Vegetation

At the bottom of canyons trees such as oaks, Arizona sycamore, and the rare Goodding ash are found. Cottonwoods and willows are also common and may spread into the lower desert, forming a ribbon of forests along permanent rivers. These narrow bands of lush vegetation along rivers are called "riparian forests." They form a rich environment that supports large numbers of plants and animals. Normally, such plants and animals could not survive in the harsh desert.

Along the sides of canyons and on the surrounding hillsides, the riparian forest gives way to oaks and, higher up, to juniper trees. Junipers can grow to 33 feet (10 m) in height. Several different species are found in the Southwest. They are especially common on the slopes around the high deserts to the north. They have a scaly bark and bluish or reddish brown cones which look like berries and are commonly called "juniper berries."

Also typical of dry hillsides are the agaves. They are especially common in the more southern deserts. Related to the lilies, they are able to store water and nutrients in thick fleshy leaves. The leaves are arranged into a rosette around a heart where the leaves join the stem. These plants flower just once, after a number of years. They produce a flowering stalk up to 6 feet (2 m) high, with clusters of flowers at the top. After flowering, they die. For humans living in the desert, the heart of the agave is a tasty vegetable, and the sap contained in the stalk can be fermented into an alcoholic beverage called "mescal."

The ecosystem of the canyons and the riverbanks is so frail that much of it has already been destroyed, due to overgrazing and to a lowering of the water level caused by excessive use by humans. Fortunately, though, some good examples of this ecosystem still exist in the Madera, Sycamore, Ramsay, and Cave Creek canyons and in the Sonoita Creek region in Southern Arizona. Most of these places face south toward Mexico and have a mild climate. For this reason, numerous species of wildlife, typical of the subtropical regions farther south, are able to exist in these northern

Opposite page: A dry streambed winds through the Canyon de Chelly in Arizona. The canyon walls, smoothed by erosion, are lined with red "desert varnish." During the dry season, the streams are often dry in the canyon, but water can usually be found a few feet underground. This makes it possible for a large number of plants and animals to survive.

A riparian forest is seen in Sabino Canyon. Oak, sycamore, ash, aspen, and willow forests are found at the bottoms of canyons.

outposts. Other similar areas are practically surrounded by the Sonoran Desert and could well be considered as oases. A magnificent example is Aravaipa Canyon. Its name comes from the Papago Indian phrase meaning "little wells." This is a very appropriate name because even during the dry season, "sinks" of water remain among the rocks and boulders within the canyon.

Fish

The last thing a person would think of looking for in a desert is fish. However, fish can be found in water holes,

rivers, and some bodies of water that almost disappear in dry periods. This might seem impossible, and so is worthy of special attention.

Moist desert areas are extremely vulnerable and so easily disturbed that some unique species of fish have probably already been lost. The loss of some native fish in the Colorado River has been a source of concern to many. Desert springs and small streams are also in danger. In the desert, water can emerge from the ground in unexpected places and with unpredictable quality. It can be alkaline or, when there is no stream flowing out, it can become extremely salty.

Freshwater springs and alkaline wetlands were once found throughout the desert. Some dried up when underground water was tapped for human use. Others have been lost, more foolishly, because livestock trampled and destroyed them. Some of these oases certainly offered refuge to unique species that have been lost forever, together with their habitats. But some of these oases still exist, such as the pools of Aravaipa Canyon and the alkaline springs of Quitobaquito in Organ Pipe Cactus National Monument in southern Arizona. Water can produce local swamps and marshes that soon fill with algae, followed by flies, gnats, and numerous aquatic invertebrates. All of these, in turn, are food for desert fish.

Over the last fifty years, many game fish and alien species have been introduced throughout the region. In many cases, the newcomers have taken over the habitat from the native species. Thus, competition from introduced species, combined with the loss of some habitats, has caused a sharp reduction in numbers of desert fish. In a few sites, the native species can still live relatively safely. For example, in the pools of Aravaipa Creek the Gila and Sonoran suckers, the longfin dace, the loach minnow, the roundtail chub, and the spike dace are still found. Some species are worthy of notice because they are adapted to almost impossible living conditions. During the dry season, the water level sharply decreases and the temperature soars. On the other hand, the rainy season can also be risky, because little streams and creeks turn into tumbling torrents.

The Sonora chub is found in only a very few sites in the Sonoran Desert. It is remarkable for its extreme endurance in the most challenging conditions. For example, about twenty years ago, a small group of them was discovered living in a tiny pool containing only a few quarts of water.

The longfin dace is much more widespread and takes advantage of every opportunity. When desert streams suddenly flood, this fish enters the rushing waters and swims for hundreds of feet and sometimes for many miles. As these streams dry up, many dace die. Some individuals, however, manage to survive in holes and depressions under boulders, where some water remains. In some streams during the dry season, water may not run during the day, but a trickle might resume at night. In this case, the longfin dace survives seeking shelter under moist debris and aquatic vegetation during the day. It then comes out at night. It requires only a few tenths of an inch of water in which to swim and eat. It feeds mainly on algae and microscopic crustaceans but may also eat rotting plant material, other types of organic matter, and occasionally small aquatic insects. It can reproduce during most of the year (usually December through July or even into September). The eggs are laid in small depressions in the streambed.

Another notable fish is the desert pupfish. At present, its range is rather limited, due mainly to a loss of habitat. Some healthy populations of this species are maintained and protected at Quitobaquito Springs in Organ Pipe Cactus National Monument and at a few other sites. This fish usually dwells in springs, as well as in marshes and backwaters. It can tolerate extreme changes in temperature, as well as variations in salt and alkali content. Most fish in desert water holes cannot tolerate such great changes. Desert pupfish living in the Salton Sea tolerate temperatures as high as 95°F (35°C) and a salt content almost three times greater than that of the sea.

Amphibians and Reptiles

Many amphibians thrive in the canyons, although they usually stay close to permanent water. The canyon tree frog, 2 inches (5 cm) long, is gray with dark speckles. It prefers to stay inside cracks and close to creeks or permanent pools. This frog can cling to vertical walls by using adhesive pads on its toes. It hardly ever moves more than a couple of leaps away from water. The call is a popping, whirring sound, lasting one to three seconds. It attaches its eggs to the underwater vegetation, either one-by-one or in clusters.

Another species found in the Southwest near permanent water is the Tarahumara frog. It is up to 4 inches (10 cm) long and is found throughout the Sierra Madre Occidental and into extreme southern Arizona. It lives near streambeds

Opposite page: The canyon tree frog, like all treefrogs, is a skilled climber and can easily climb smooth vertical walls, thanks to special adhesive cushions on its toes.

The Sonora Mountain king snake, also called the "false coral snake," is a completely harmless reptile, without fangs and poison glands. This is in spite of its extraordinary resemblance to the highly poisonous coral snake. The camouflage used by a harmless animal to resemble a dangerous one is known as "mimicry."

in mixed forests of oaks, willows, and sycamores. It never goes much more than 6 feet (2 m) away from the stream bank. Although these amphibians prefer running water, they will gather near quiet pools during the dry season. Little is known about their breeding habits, but they probably lay their eggs during the summer rains.

Many of the desert reptiles, including rattlesnakes and lizards, have already been discussed. Some of these are also found in the canyons, while different species live in the lower deserts and in the transition zones at higher elevations. An example of the second group is the tiny coral snake, which is hardly ever longer than 16 inches (40 cm). It is a beautiful snake, with a series of black, yellow, and red bands. Despite its attractive colors, this species is extremely poisonous. Due to its shy habits and its solitary life, and also because of its small size and tiny fangs, this snake seldom inflicts severe bites on people. Nevertheless, it is certainly not a good idea to handle it if you are inexperienced with snakes.

At higher altitudes, the Sonora Mountain king snake is found. At first glance, this snake, which is harmless to peo-

The brown vine snake, with a large and elongated head on a long and very slender body, belongs to a group of snakes with poison-bearing teeth that are placed far back on the upper jaw. The bite of these snakes is generally not dangerous to humans but is perfectly effective for paralyzing small animals on which this thin snake feeds.

ple, looks alarmingly like a coral snake, but much larger. It has broad red bands separated by narrow bands of yellow and black. This snake is nocturnal and preys on other snakes, lizards, and small mammals. When it is threatened, it reacts by striking in the same manner as a poisonous snake. Thus this king snake mimics the behavior, as well as the coloration, of the more dangerous species. Actually, the Sonoran Mountain king snake kills its preys by constriction, crushing its victim to death in its coils.

The brown vine snake lives mainly in Mexican canyons, especially where there are streams bordered by oaks and sycamores and hillsides covered with bushes. It is rare in the United States, being found only in the Pajarito Mountains in southernmost Arizona. This snake is about 4 feet (1.2 m) long and extremely thin—often no more than the diameter of a pencil. It is a grayish or yellow-brown color and has a long head with a pointed nose. The vine snake is an excellent climber that hunts for lizards both in trees and on the ground. It kills its prey by injecting poison through grooves in enlarged teeth located toward the back of the upper jaw.

BIRDS OF THE RIPARIAN FORESTS

In the desert, few experiences are more inspiring than a stroll through a canyon early in the morning or at dusk. The air is filled with the buzzing of insects—ringing metallic sounds, almost electric. Butterflies drift by under the leafy branches of sycamores, oaks, and cottonwoods. Beneath the trees, close to the water's edge, cooler air and higher humidity give relief from the scorching air of the nearby open desert.

The black phoebe, a flycatcher with a handsome black-and-white plumage, patrols its territory along rock-strewn creeks, busily catching insects in flight. Many lizards dart among rocks spotted with ferns and agaves. This is a rare instance where ferns and agave plants grow so close together. From the trees overhead can be heard the almost shrieking call of the sulphur-bellied flycatcher. Hummingbirds zip around, frantically looking for flowers and nectar, while a quiet tree frog rests in its crevice waiting for its prey to come within reach.

Birds of the Trees and Meadows

The acorn woodpecker is often discovered while exploring. Its plumage looks like a red, black, and white uniform; its back is blue-black, its chin is also black, and its rump is white. White on the upper breast extends around the side of the head and joins just above the bill. It has a bright red crown and yellow eyes. This bird lives in complex social groups, in which cooperation occurs during breeding. One male and one female will reproduce, while the rest of the group helps to find food for the nestlings. Acorn woodpeckers got their name because of their habit of storing acorns in small holes which they peck in trunks or stumps.

Flocks of the harlequin quail search in the grassy meadows under oaks and along the sides of the canyon for seeds, green shoots, and insects. Males are striking, with black-and-white heads that resemble the painted face of a clown. They have streaked backs and maroon bellies, with white spots on the sides. These birds almost seem tame. When disturbed, they prefer to squat and hide rather than to fly away. But when an intruder gets too close, they suddenly burst up, flapping their wings and flying a few feet.

Daytime Birds of Prey

The black hawk can be found in areas where streams flow or marshes form. This bird's numbers have sharply

Opposite page: Lush forests, as shown in this picture taken in Arizona, form along the banks of rivers enclosed by high walls in western canyons. These environments teem with life and are favorite sites for bird-watchers.

A flock of black vultures roosts on a tree. The vultures of North America are not related to those of Europe, Asia, and Africa. In the United States, three species in the vulture family are found. Two of them are rather frequent (the black vulture and the turkey vulture), while the third, the California condor, is on the verge of extinction.

declined in the United States due to destruction of its favorite habitats. It is a beautiful bird of prey, with a wingspan of over 3 feet (1 m). It is mostly black, with long yellow legs and a broad white band across the middle of its tail. It prefers riparian forests and moist environments where it can hunt for frogs, toads, crawfish, and reptiles.

Smaller than the black hawk is the gray hawk, which has a wingspan of usually less than 3 feet (1 m). It is found in many of the same environments as the black hawk, but it also hunts in drier areas. The gray hawk preys mainly on lizards. Actually, both species are more typical of tropical

climates, and the southern canyons of Arizona are the northernmost point of their range.

A familiar sight in desert skies is the soaring circular flight of vultures, always in search of carrion. There are two common species of vultures. The first is the turkey vulture, with a wingspan of 6 feet (2 m). Its main features are bare red skin on the head and "two-toned" black-and-white wings. The second species, the black vulture, is smaller, with a wingspan of about 5 feet (1.5 m). It has a short, square tail, a black head, and white patches at the tip of the spread wings. Both species feed almost exclusively on carrion, although black vultures also visit garbage dumps outside of towns. They seem to be effortless fliers, banking into turns and soaring on rising currents of hot air. The turkey vulture seems to soar constantly, with hardly ever a beat of its broad wings. The black vulture, on the other hand, can be identified at a great distance by its more frequent flapping. Both species lay one to three spotted eggs in a nest built in a tree, a hollow log, or a cave, or sometimes even on the ground.

Still another dweller of the southern deserts is the zone-tailed hawk. It can sometimes be confused with the turkey vulture because of its short, broad tail, the proportions of its wings, and its soaring flight. Despite its resemblance to a vulture, this bird is a real hawk.

Other Birds

The wooded canyons and riparian forests provide a refuge for many other species of birds. One of them is Scott's oriole, a brilliant yellow-and-black bird. It is common throughout the region, especially in yucca forests and in junipers and oaks. It feeds mainly on fruits and insects, and its call is similar to that of the western meadowlark.

Along the canyon walls, it is common to hear the haunting song of the canyon wren. It has a rusty back and tail and a white upper breast. Higher up the canyon, a low, coarse, croaking sound like the call of a female turkey might be heard. The elegant coppery-tailed trogon, which can be common in the canyons in some years, can be spotted here. It has a rosy belly and a long, copper-colored tail. It feeds on fruit and insects and builds its nest in tree holes, where it lays three to four eggs.

The Wonderful Hummingbirds

No description of the birds living in the deserts of the Southwest would be complete without a few words about

A Costa's hummingbird rests on the tiny nest it has built in a tangle of branches of a thorny bush. More than a dozen species of hummingbirds live in the western part of the United States, but this is just a small sampling of the hundreds of species found farther south. There hummingbirds find abundant habitats in Mexican forests and on mountain highlands.

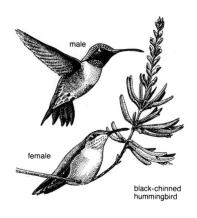

black-chinned hummingbird

broad-billed hummingbird

blue-throated hummingbird

Rivoli's hummingbird

the hummingbirds. These birds are fireballs of pure energy as they zoom up and down the canyons. Sometimes they venture into the lower deserts or may even reach high mountain meadows. Over fifteen species of hummingbirds have been described in southern Arizona, New Mexico, and Texas. During a one-day outing, visitors may well come across five or six different species, especially in those canyons where special feeders containing sugar water have been installed.

Hummingbirds are famous for their flying skills, made possible by a specialized wing structure. The bones in a wing correspond to those of an arm and hand. Keeping this in mind, the structure of a hummingbird's wing can be understood. The bones of the "forearm" are greatly reduced, so most of the wing is supported by the bones of the "fingers," which are long and fused to form a single bone. The bones of the "upper arm" are short and stubby, and the exceptionally strong muscles of the breast are attached to them. These muscles are used for flying. This structure enables the wings to be flapped at a prodigious rate, up to ten times a second or faster. The rapid wing motion produces a characteristic humming sound. Hummingbirds can also flap from their "wrists." This allows great flexibility of wing movement. The wings can be moved in circles, allowing hummingbirds to fly straight up or down, or even backwards. They can hover in front of flowers while they sip nectar. They can back up, move side-to-side, or go up or down to the next flower. Their method of flight is also helpful when hunting small insects. They simply dart back and forth, picking off tiny gnats in a swarm.

The plumage of hummingbirds is different from that of other birds. Many of their bright colors are not the result of pigments, but instead are due to the structure of the feathers. Hummingbird feathers refract (bend) light, much the way that small drops of water in the atmosphere refract light to create a rainbow. Several of the hummingbird's bright reds, blues, greens, and purples originate in this way. When seen at certain angles, these colors vanish, and the bird appears black or dull gray. If, for example, a male bird with a gorgeous red throat turns its head a little, the bright red will become black, only to reappear with another slight movement of the head.

Several species of hummingbirds are common in this region. In the lower desert and at the outflow of canyons, there may live the tiny Costa's hummingbird, barely 3

screech owl

whiskered owl

flammulated owl

ferruginous owl

inches (7 cm) long and weighing a few tenths of an ounce. The male has an iridescent green back. Its head, throat, and side feathers sticking out from the cheeks are a bright amethyst. Farther up the canyons and at higher elevations, the black-chinned hummingbird, about the same size as Costa's hummingbird, is often found. This bird also has a green back, but its throat is black, underlined by a purple band that extends to the top of the breast. Higher still can be found the broad-billed hummingbird, which is easily recognized by its dark green body, bright orange bill, and forked tail. The large blue-throated hummingbird can be 5 inches (13 cm) long and is recognized by its broad tail with white patches in each corner. Its back is greenish, its throat is an iridescent blue, and its face is marked with a double white streak. The striking Rivoli's hummingbird is also about the same size and is sometimes referred to as the "magnificent hummingbird." The male is uniformly dark except for a glistening green throat and a violet patch on the crown. All hummingbird species build tiny nests with such materials as lichens, wool, and spiderwebs. They usually lay two white eggs at a time.

Nighttime Birds Of Prey

At night in the canyons, several species of birds of prey come out to hunt. Moving up the canyons, different birds populate the different habitats. The elf owl and the screech owl prefer the lower desert. Higher up, where oak trees grow, the whiskered owl is more likely to be encountered. It is easy to tell this owl apart from the screech owl by its hooting call, sounding like a Morse code signal. Higher up, where pines begin to appear, the small flammulated owl can be found. This is a tiny, insect-eating species that throws its low-pitched hoot like a ventriloquist. This makes it difficult to tell where the sound is coming from, so it is not easy to spot the bird even when it is only a few feet away. Forests along river bottoms are sometimes the habitat for the ferruginous owl, a southern relative of the northern pygmy owl. This bird just barely crosses the Mexican border into southern Arizona. Higher still, above the desert zones, there are other owl species. Only one species, the great horned owl, occurs in all the zones.

All of the nocturnal birds mentioned above are birds of prey and hunt small rodents, birds, and reptiles. Some, though, such as the elf owl and the flammulated owl, are mainly insect-eaters.

MAMMALS IN THE CANYONS

Numerous species of mammals live in the canyons and in the transition zones of the North American deserts. Many of them live in both the low and high deserts. Groups of peccaries roam the hilly slopes and canyon bottoms. At higher elevations white-tailed deer are found. Mule deer move between the desert floor and the lower reaches of the canyons. Throughout the canyons are various squirrels and chipmunks.

Bighorn Sheep

The canyon walls, rocky outcrops, buttes, and mesas are ideal habitats for the desert bighorn sheep. A ram can be up to 4 feet (1.25 m) long and weigh 176 pounds (80 kg). Females are about one-third smaller. This species is smaller than its northern relatives, with a thinner coat and a less distinct white patch on its rump. The fur of these sheep is a brownish yellow-gray. Rams have massive curved horns that keep growing throughout life. Females' horns are much smaller and less curved.

The desert bighorn sheep has sharp eyesight but not very good hearing or sense of smell. Its hooves are cloven and concave on the sole, which probably helps soften landings on rocks when it jumps from boulder to boulder with leaps of up to 20 feet (6 m). The shape of the hoof is also important for good ground-gripping ability. These animals can climb almost vertical rock walls. This ability helps to reduce the losses of bighorn sheep to predators.

Desert bighorns feed on grass, but when grass is not available they also eat cacti, seeds, and shrubs. They usually travel to water holes at least once a day to drink, but it is possible for some individuals to go without drinking water for five days or more. For most of the year, bighorns live in flocks composed completely of members of the same sex. The rams tend to stay apart from the ewes and lambs. In autumn, the breeding season begins, and the rams begin to fight to determine their dominance status within the group. At this time, the loud crack of clashing horns can be heard as far as a mile away. The dominant rams group their females into harems and do most of the breeding.

Squirrels And Chipmunks

In the oak, aspen, and sycamore woods is found the gray Arizona squirrel. This animal feeds mainly on acorns, seeds, and other plant matter. On the canyon floor and along steep cliffs, lives the cliff chipmunk. It is easy to spot,

Opposite page: The desert bighorn sheep is a subspecies of the Rocky Mountain bighorn sheep. This variety lives in the dry areas of the United States. It thrives in the desert, while its relatives typically dwell in mountain areas, especially in alpine meadows above the tree line, from Alaska down to the Mexican Sierras.

as it runs swiftly from rock to rock and among the bushes. It is 9 inches (23 cm) long and its grayish fur becomes brownish red toward the front of its body. A typical black-and-tan stripe runs along its sides.

There are many species of chipmunks in North America, but the cliff chipmunk is the only one that lives in the true desert. All the other species in the Southwest are found at much higher elevations in pine and fir forests. The cliff chipmunk sometimes even ventures into the open desert, where it builds a nest under a pile of stones or in a crack in the rocks. It does need to drink, but during the dry season it can survive long periods without water. It gets the moisture it needs from cacti, berries, bulbs, and occasionally also insects. But grass, seeds, acorns, and other nuts are also a part of its diet and often are stored in chambers near its nest. Unlike its relatives living farther north, this species probably does not hibernate, but is active all year. During cold periods, it feeds mainly on food it stored earlier.

Small Carnivores

One of the most interesting animals of the southern canyons is the coati, a member of the raccoon family. It can

North American gray squirrels feed mainly on material that they collect in trees. Unlike red squirrels, they adapt easily to different environments and are not bothered by the presence of humans.

A coati is caught off guard by a photographer. This small carnivore, a member of the raccoon family, is widely distributed all over tropical America and has been living in the United States for a few decades. It originated in Mexico and has spread to Arizona, New Mexico, and Texas.

grow to 3 feet (1 m) in length, of which about half is tail. The tail has light and dark brown rings, and it is held straight up as the animal walks about. Its fur is brown, and its long snout has a disk-shaped structure at the tip, similar to a pig's. This is used to dig and root in the soil. This small mammal is a plantigrade, which means that its body weight is evenly distributed on the soles of its feet and toes.

Coatis communicate among themselves with a series of low grunts and odd twittering sounds, which resemble the call of a bird. They are highly social and move around in bands of up to twenty members. Sometimes, though, isolated individuals can be found. Coatis are omnivorous, and their diet is quite varied. They eat insects and other invertebrates, seeds, fruits, berries, and at times, even small lizards, snakes, rodents, and bird eggs.

Coatis are excellent climbers, especially when being chased by a predator. If they are cornered, they put up a fierce fight, using long front claws as weapons. As a result, they have only a few enemies in their natural environment. The young, generally four to six, are born in July. Females leave the main group to give birth. But as soon as the young

An alert ringtail poises for action. This peculiar animal might at first sight resemble a raccoon, but it is smaller and more nimble. It can climb trees swiftly, and this is the main reason for its common but inappropriate name: "ring-tailed cat."

are weaned and can move about on their own, they and their mothers rejoin a nearby group.

A common sight in the Southwest deserts is a group of coatis waddling across a canyon with their tails held up straight. They often stop and stand on their hind legs, using their tails as a support, and sniff the air.

Oddly enough, coatis have been moving north. One hundred years ago, they were completely unknown in the United States. Then, some individuals made their way north from Mexico, and their offspring are now well established in Arizona, New Mexico, and Texas, but still in areas close to the Mexican border.

Pictured are: a powerful jaguar *(left)*, a margay on a trunk, and a jaguarundi. These cats were once common in the American deserts, but today they live almost exclusively in Mexico and farther south. They have been hunted both by ranchers worried about possible damage to their livestock and by fur hunters.

At night, several other animals venture out. Among them, the ringtail is fairly common. It is found from the bottoms of canyons all the way up to higher elevations where oaks and junipers grow (from Mexico to Nevada and Utah). This peculiar animal is considered by some scientists to belong to the raccoon family, while others place it in a family of its own. Despite its nickname, the "ring-tailed cat," it is not related to cats. A slender animal, the ringtail is 30 inches (75 cm) long and weighs about 2 pounds (1 kg). It has a long tail with prominent black-and-white rings. Its legs are short, and its feet are catlike. Its head resembles that of a fox, with huge eyes and large rounded ears. The ringtail is a nocturnal animal that spends the day hidden in a cave or the hollow of a tree. It is basically a shy creature, but it can be seen at dusk as it looks for food along steep slopes or in trees. Its prey are insects, larvae, and rodents. The young, three to four per litter, are born in May or June and have a snowy white coat. They become independent around three months of age.

The Cats

The cat family is one of the major groups of predatory mammals. At least six species live in the Southwest deserts. The largest and most powerful is the jaguar. This magnificent animal is still widely distributed in South America, and it once lived as far north as Colorado. At present, its numbers are sharply reduced. It is considered extinct in the United States but still can be found in the wildest areas of Mexico. Jaguars are large, heavy cats which often grow 8 feet (2.4 m) long and weigh 220 pounds (100 kg). The jaguar's

yellow or tan fur has a pattern of distinct "rosettes" with a dark center surrounded by a paler black ring. This pattern is quite different from the spots of leopards found in Africa and Asia. When the jaguar hunts, it does not chase its prey with sudden bursts of speed. Instead, it lies in ambush and, at the right moment, jumps on its victim's back bringing it down by sheer force of weight. It prefers peccaries, but on occasion it will also hunt deer or livestock. The extinction of the jaguar in the northern parts of its range has been a serious loss for the entire ecosystem.

This beautiful predator deserves a place in the Southwest deserts once more, but this seems highly unlikely. Any attempt to reintroduce the jaguar would likely cause a strong reaction from local ranchers, worried about the consequences to their livestock.

Another large carnivore, still quite common, is the

The mountain lion is the member of the cat family most widely distributed in North America. It lives from the northernmost regions to the extreme southern point on the continent. It can still find refuge from humans in remote areas. It has been so heavily hunted, though, that its populations have been greatly reduced. Mountain lions have disappeared altogether in many eastern regions where they were once common.

mountain lion or cougar. Growing 6 feet (2 m) long and weighing nearly 200 pounds (90 kg), the mountain lion is slightly smaller than the jaguar. It ranges from northern North America to the tip of South America. At one time, it lived both on the Pacific and on the Atlantic coasts. Today, it no longer exists in the Atlantic regions.

The fur of the mountain lion is usually gray or light brown, without spots or other markings of any kind. Although more slender than the jaguar, it is nevertheless a powerful predator. It feeds primarily on deer but also hunts peccaries and smaller game. It hunts by stalking its prey, then leaping up to 20 feet (6 m) onto its victim. It may also lie in ambush and wait for the right moment to seize its prey, as does the jaguar. Its favorite habitats are the higher areas, canyons, and hillsides, but it will occasionally come down to the desert floor.

The bobcat, or lynx, is an extraordinarily elusive animal, and it is difficult to see one even in areas where it is still rather common.

Each mountain lion has its own territory, which can be hundreds of square miles large. It marks the territorial boundaries by urinating on the soil and on dry vegetation. Breeding can occur at any time of the year. The female usually gives birth to two young. Dark spots on the coats of juvenile animals disappear as they near adulthood.

The other species of desert cats are much smaller. The most common small cat is the bobcat. It is found throughout the region but is extremely shy and thus very hard to spot. This animal is about 3 feet (90 cm) long and can weigh 40 pounds (18 kg). Its fur is generally light brown or yellow-brown, with numerous small dark spots. Its tail is short, with

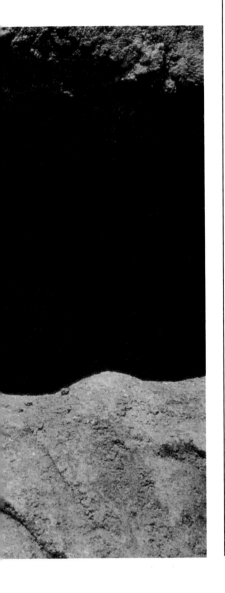

darker rings. Its ears are large and pointed, tipped with a typical tuft of dark hairs.

Large paws, armed with strong claws, make this animal a formidable predator. Its victims are rodents, jackrabbits, young deer, ground-dwelling birds, and even large insects. Mostly solitary and nocturnal, the bobcat hunts by stealth. It waits patiently, then suddenly jumps out from its hiding place, seizes its prey, and overpowers it. The young are born at any time of the year and look like domestic kittens. They are, however, very ferocious if disturbed. Unlike kittens, they have larger paws, stubby tails, and tufted ears.

Unfortunately, the other species of cats are quite rare in the region, except perhaps in Mexico. Two very similar species can occasionally be found: the ocelot and the small margay. The margay is really a forest species and barely reaches the edge of the desert in southern Texas. Both species have a yellow-brown coat dotted with darker rosettes (like the jaguar) arranged in broken lines. The ocelot hunts on the ground and in trees. It is extremely nimble and feeds on rodents, birds, large insects, and sometimes reptiles. The fur of these two cats is highly prized by hunters and, in the past, by the fashion industry. Heavy trapping has thus contributed to a sharp decrease in numbers of these species.

The sixth species, the jaguarundi, is also rare in the United States but can still be found in Mexico. It has dark fur and is about twice the size of a domestic cat. It has a long body and actually resembles a large weasel more than a cat. It hunts mainly at night for birds, small mammals, and large insects. The jaguarundi prefers brush areas and woodlands, unlike the ocelot, which can be found in rocky gullies and in open areas.

All of these cats cause only minimal damage in agricultural regions, and the smaller ones are no threat to farm animals. Despite this, these animals still fall victim to commercial trappers seeking precious furs. The Southwest deserts could still support these magnificent creatures, if given the chance.

GUIDE TO AREAS OF NATURAL INTEREST

The southwestern regions of the United States and desert areas in adjacent Mexico are particularly rich in parks, forests, national monuments, wilderness areas, and wildlife refuges. In the United States, all of these areas are open to the public and most are easily accessible on paved roads. Some remote areas require access by hiking or with a four-wheel-drive vehicle. Most areas have campsites, and motels and restaurants are usually found in nearby towns. In Mexico, on the other hand, nearly all areas of natural interest are much harder to reach. The Mexican states of Baja California, Sonora, and Chihuahua contain vast wilderness areas. Here the roads are often in poor repair, where they exist at all.

The national parks and wilderness areas described below are truly rugged and wild environments. Visitors must be prepared for bad weather and any other problems that might arise. Modern facilities may be lacking. When planning long hikes in uninhabited areas, it is absolutely necessary to have a compass and a good topographic map. In national parks, most trails have registration stations where visitors can sign in before a hike and sign out when they return. Some parks require all hikers to register before a trip. It must be kept in mind that some areas are so remote that no clearly marked trails can be found, and many regions are not patrolled by rangers.

Always carry an adequate supply of drinking water when hiking in desert areas. On the desert floor, daytime temperatures can exceed 120°F (49°C) in the summer, and there might not be any water along the trails for many miles. Avoid hiking in the open during the hottest hours of the day. Emergency food, a flashlight, and a first aid kit are helpful for coping with unexpected situations. In the unlikely event that you are bitten by a snake, contact a doctor as soon as possible.

Normally, camping is allowed throughout the region, but in exceptionally dry years, it might be restricted in some areas because of the high risk of fire. In the desert, it is a good idea to sleep inside a tent, even when nights are comfortably warm. Scorpions and snakes come out at night and might crawl into a sleeping bag lying in the open. A tent with a protective net on the entrance will prevent any such incident. Never camp in dry riverbeds. Even when the sky looks clear, it is always possible that a local storm, perhaps many miles away, will send a raging torrent of water and debris down the river without warning.

Opposite page: The Anza-Borrego Desert in California is viewed from the air. When spring rains come, this arid stretch of land turns into a richly-colored carpet of flowers.

Below: This illustration indicates the American desert area. It covers the southwestern United States and bordering regions in Mexico.

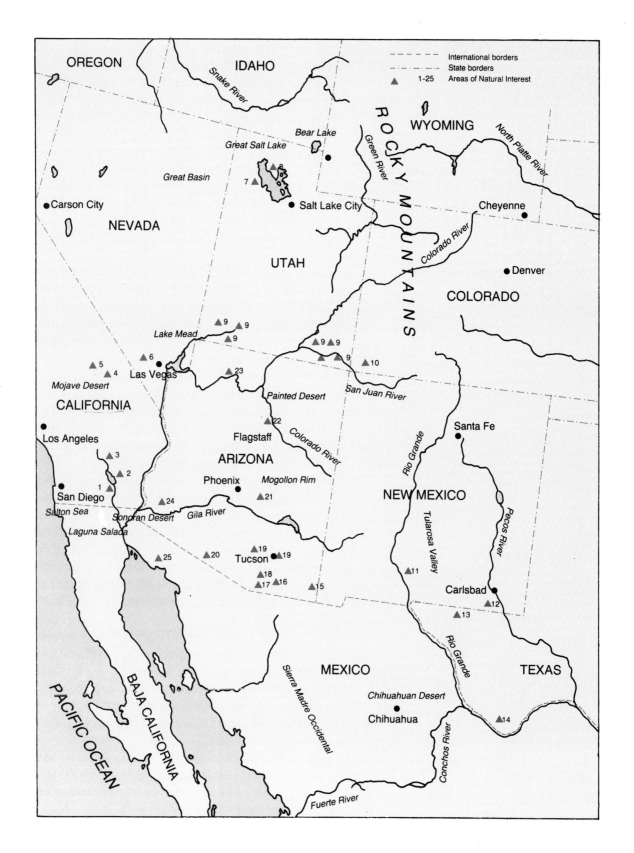

UNITED STATES

California:
Anza-Borrego Desert State Park (1)

This is a large park, about 580 sq. miles (1,500 sq. km). It can be reached by Interstate 8 east from San Diego or by Route 78 east from Escondido. Routes S2, S22, and S3 cross other regions of the park.

The area is famous for the display of flowers in spring, especially after heavy rains. At such times, this state park can be crowded. The sand dunes are especially attractive to people with all-terrain vehicles. As a result, wildlife can be hard to find amid all the noise and disturbance. A careful search in quieter places, though, might lead to more rewarding results. On the sand dunes, the sidewinder rattlesnake can be found.

California:
Salton Sea (2)

This large salt lake can be reached by Route 86 south of Interstate 10 (near Coachella). Route 111 borders the east side of the "sea."

The lake is shallow and lies in a depression between stark desert mountains and ridges. The resulting contrast is truly impressive.

Many desert animals live in the surrounding brush. Also, aquatic animals dwell in the reeds and swamps at the water's edge. Wading birds are seen frequently, especially in winter. Large numbers of black-necked stilts and marbled godwits are often encountered. Occasionally, it is possible to spot a skimmer.

California:
Joshua Tree National Monument (3)

This national monument consists of about 850 sq. miles (2,200 sq. km) of high desert (part of the Mojave Desert). It is reached via Interstate 10 east of Los Angeles.

The main attraction in the area are the Joshua trees, but many creatures abound here, too. For example, over 260 bird species have been recorded. The bighorn desert sheep also is found here. Tourist facilities are scarce, and much of the park is completely wild.

California:
Death Valley National Monument (4)

This huge national monument covers nearly 2,900 sq. miles (7,500 sq. km). To get to Death Valley, take Route 395 north from Interstate 15 (just north of San Bernardino) to Olancha. Then turn east on Route 190 to the park. A winter visit is advisable, since the temperature can get unbearable during the summer.

Within the valley are the most desolate stretches of desert on the entire continent, as well as the lowest at 282 feet (86 m) below sea level, and hottest points. Despite the hostile environment, about six hundred species of plants

Opposite page: The map shows the locations of the main areas of natural interest in the southwestern United States and adjacent Mexico.

The Great Salt Lake, when seen from a distance, seems unreal, with blue, gray-green, and white reflections. It is about 30 miles (50 km) wide and 60 miles (100 km) long. Its maximum depth, however, is only 33 feet (10 m) and varies greatly depending on rainfall and meltwater flowing in from the surrounding mountains.

California:
Owens Valley (5)

Nevada:
Las Vegas region (6)

Utah:
Great Salt Lake (7)

Utah:
Bear River (8)

Utah:
Southern canyons, plateaus, and mesas (9)

and a wealth of animals, including bighorn sheep and coyotes, live there.

The main part of this valley is about 60 miles (100 km) long and can be reached via Route 395 north of the turn-off to Death Valley. This is high desert, mostly of the Mojave type. It offers an interesting contrast to the southern deserts, which are generally warmer in winter.

Famous for its other tourist attractions, Las Vegas is also an excellent starting point for excursions into the Great Basin deserts of Nevada. The Desert National Wildlife Refuge, as well as vast areas covered with creosote bush and sagebrush, are close at hand.

Visitors are likely to spot bighorn sheep, pronghorn, golden eagles, sage grouse, prairie falcons, and numerous reptiles. When spring rains come, the region blossoms with brightly-colored flowers.

The Great Salt Lake is about 60 miles (100 km) long and up to 31 miles (50 km) wide. It cannot be reached on its western side, but along the eastern and northern sides are several state parks within easy reach of Salt Lake City and Odgen.

The lake is a huge landlocked sea surrounded by high mountains. Much of the land around the lake is covered with sagebrush and is rich in animal life. Cormorants and pelicans can be seen on the lake.

This is a refuge for migratory birds, covering about 70 sq. miles (180 sq. km) in a swampy area on the northeastern side of Great Salt Lake. Areas around the edges of the refuge can be entered from west of Brigham City.

This area is especially interesting during spring and fall migrations. At these times, impressive numbers of migratory birds can be observed here. Among them are blue-winged teals, cinnamon teals, many species of geese (especially Canada geese), and sandhill cranes.

The numerous national parks and monuments in southern Utah are spectacular with spires, gothic columns, and finely-carved natural arches. In other areas, the wide open spaces are dotted with buttes and mesas. These gigantic chunks of rock jut suddenly out of the plateaus. The geological formations are fascinating, but no less so than the plants and animals. Junipers and oaks are widespread. Cottonwood trees grow around pools of permanent water. These areas provide food and shelter for many creatures.

Among the common birds are rock wrens, canyon wrens, and black phoebes. On the exposed plateaus the pronghorn, sage grouse, black-throated sparrows, and collared lizards make their homes. This entire area deserves an extended visit to fully appreciate the variety and beauty of its natural wonders.

Colorado: Mesa Verde National Park (10)

This national park covers about 46 sq. miles (120 sq. km) on the border of the arid Southwest regions. It can be reached by road south from Route 160, about 8 miles (13 km) east of Cortez.

The park is worth a visit for its spectacular canyon scenery. There are animals and plants typical of the higher areas of the Sonoran Desert, as well as of the Rocky Mountains.

New Mexico: White Sands National Monument (11)

This national monument covers about 154 sq. miles (400 sq. km). It can be reached by Highway 70/82 between Las Cruces and Alamogordo. There is little access to the park except by foot.

This area, famous for its wind-swept dunes of pure white gypsum crystals, is especially interesting geologically. Besides the unique gypsum dunes, there are numerous signs of volcanic activity, including recent lava flows.

New Mexico: Carlsbad Caverns (12)

This extraordinary area offers 21 sq. miles (55 sq. km) of caves, created by the action of water inside a huge limestone reef complex. Some caves have been explored to depths of nearly 1,000 feet (300 m), and certainly they go even deeper. Inside these caves, the temperature is constant throughout the year, around 56°F (13°C). The park organizes guided tours of the most spectacular caves.

Another attraction in summer is the amazing emergence at dusk of the Mexican free-tailed bats from their roosts deep in the caverns. They erupt by the thousands and pour back into the depths at dawn. In some summers, as many as a million bats roost in a single cave!

Texas: Guadalupe Mountains National Park (13)

This park includes about 27 sq. miles (70 sq. km) of mountains and Chihuahuan Desert. This mountain range, a northern outpost of the Sierra Madre of Mexico, is reached via Route 62/180 east of El Paso.

Canyonlands National Park in Utah is seen near the end of winter. In the background, peaks of the Rocky Mountain range are still topped with snow.

**Texas:
Big Bend
National Park (14)**

The area is interesting because it is in a region where the fauna and flora of the Chihuahuan Desert overlap with those of the Rocky Mountains and of eastern America. Typical animals are elk, mule deer, turkeys, raccoons, coyotes, and bobcats.

A land of contrasts, this fabulous national park lies north of the Rio Grande and the Mexican border. It covers about 193 sq. miles (500 sq. km) and includes mountains and low desert basins. Access is via Route 118 or Route 385 south from Interstate 10.

Within the park are desolate flats that sizzle in the hot sun and much cooler highlands that contain wide stretches of grasslands and forests. Coyotes, mountain lions, peccaries, and various reptiles and birds live here. In the harshest parts of the lower desert, the resurrection plant can be found.

**Arizona:
Chiricahua
National Monument (15)**

This small national monument covers only 4 sq. miles (10 sq. km). It is part of the Coronado National Forest in the far southeastern part of Arizona.

The park consists of extraordinary spires and canyons of volcanic origin. There is a lot to see, both here and in the nearby Chiricahua Mountains. From Portal, 19 miles (30 km) to the east, it is possible to enter Cave Creek Canyon, which leads to Rustler Park near the mountain crest. The feel of the region is decidedly Mexican. At higher elevations, look for Apache pines (with extremely long needles), buff-breasted flycatchers, and Mexican juncos. With a bit of luck, a mountain lion might be seen.

**Arizona:
Ramsay Canyon (16)**

This canyon in southeastern Arizona is certainly worth a visit. To get there, take Route 90 south from Interstate 10 (just west of Benson). Go to Huachuca City and Sierra Vista and then go south on Route 92. Ramsay Canyon is off to the west in the Huachuca Mountains.

The animals and vegetation here are typical of the upper Sonoran zone. The main attraction, however, is the Mile Hi Ranch within the canyon. Here, many bird feeders containing sugar water have been installed. The feeders attract a great many hummingbirds. During the summer, six species or more can be seen at the site. With much luck, rarer species of hummingbirds might be observed. Among them there are the violet-crowned and white-eared hummingbird as well as the blue-throated and Rivoli's hummingbird. This is also a very good spot to see local orioles and trogons.

Below the mouth of the canyon is the valley of the San Pedro River. Most of the valley floor is cultivated, but in the riparian forests some interesting species, such as the Mexican duck (now considered a race of mallard), can still be found. In the fields, the scaled quail and peccaries might be seen.

Arizona: Patagonia's Sonoita Creek (17)

This is a region of upper Sonoran zone close to the Mexican border. Take Route 82 northeast from Nogales for about 18 miles (30 km). Just before the town of Patagonia is a small wildlife refuge.

A hike through this refuge and inside the canyon can be a delightful experience. Tall cottonwoods and sycamores line the creek, and oak trees abound. Gray and black hawks, bobcats, roadrunners, and many other animals can be found. In recent years, the thick-billed kingbird and rose-throated becard have been seen regularly. In small side valleys, the five-striped sparrow might be spotted.

Arizona: Madera Canyon (18)

This is a typical southeastern Arizona canyon, located in the so-called Mexican Mountains. It is part of the Coronado National Forest, and efforts have been made recently to preserve the canyon for the future.

Drive south from Tucson on Interstate 19 to the town of Continental. Then head east on a dirt road, following the road signs, for about 12 miles (20 km) to the canyon. Below the canyon, wide stretches of desert grassland contain several interesting species. Unfortunately, this area is heavily overgrazed but still attracts large flocks of sparrows and lark buntings in the winter.

When entering the canyon, oak trees and Arizona sycamores are encountered. Higher up are junipers, ponderosa pines, and Douglas firs. The creek has water throughout most of the year, so the wildlife in this enchanting canyon is abundant and varied. Found are ringtails, bears, coatis, squirrels, and chipmunks. Among the most frequently seen birds are hummingbirds, flycatchers, and warblers. In grassy clearings are harlequin quail, and at higher elevations the coppery-tailed trogon can be seen in summer. In the trees, the sulphur-bellied flycatcher, the Arizona woodpecker, and the acorn woodpecker might be spotted. Among the nocturnal birds of prey, look for the elf owl and whiskered owl.

Many other canyons in this area are probably just as spectacular, but they are much harder to reach. In some cases, they are privately owned. However, it is difficult to believe that Madera Canyon won't be able to satisfy its visitors.

Arizona: Saguaro National Monument (19)

This national monument has two sections. The Rincon Mountain section covers about 73 sq. miles (190 sq. km) and is located 17 miles (27 km) east of Tucson on East Broadway. The Tucson Mountain section covers 48 sq. miles (125 sq. km) and is located 15 miles (24 km) west of Tucson on Speedway Boulevard.

The Rincon Mountain section has a limited area of lower desert where saguaro cacti can be found. The rest is at higher elevations in the upper Sonoran zone. Hiking here is excellent. Look for numerous birds. Moreover, bears and ringtails might be spotted. More common are squirrels and other rodents.

The Tucson Mountain section contains a spectacular forest of giant saguaros. Among the numerous other plants are barrel cacti, ocotillo, and paloverde. This area is home to foxes, ground squirrels, cactus wrens, thrashers, black-throated sparrows, Gila woodpeckers, and many other vertebrate species. At night, look for tarantulas, scorpions, rattlesnakes, and possibly even a Gila monster. This is a delightful place, and all visitors will enjoy it. Moreover, it is close at hand, just a few miles from Tucson.

When in the Tucson Mountain section, visitors should tour the Arizona-Sonora Desert Museum, located at the edge of the monument. It contains a superb collection of plants and animals from the Sonoran Desert. It is an ideal place to become acquainted with the natural history of the region before setting off on your own desert adventures.

Arizona: Organ Pipe Cactus National Monument (20)

This is one of the most beautiful national monuments in the Southwest. It covers some 500 sq. miles (1,300 sq. km) mostly in the lower Sonoran zone. It can be reached by Route 85 south from Interstate 8 and through the towns of Ajo and Why. From Tucson take Route 86, then turn south on Route 85.

The main attraction here is the undisturbed desert with magnificent stands of organ pipe and saguaros cacti. The senita, another large cactus, can also be found here, together with most of the typical desert plants.

An excellent dirt road makes a loop through the mon-

Opposite page: The landscape of a desert floor in the Cedros Islands off the eastern coast of Baja California is dotted with the tall stalks of agave. This plant is typical of this desert region.

ument. Many short trails branch from it, all extremely interesting to explore. Various types of animals live in the area, such as peccaries, mule deer, foxes, coyotes, ground squirrels, kangaroo rats, many birds, and other vertebrates.

Some time should be spent visiting Quitobaquito Springs, located inside the park. This is one of the few water holes in the area. Although these springs have rather alkaline water, they are home to one of the last remaining populations of desert pupfish. These tiny fish can be seen in the sink holes. Larger ponds are often visited by birds that fly north from Mexico.

Perhaps because it is so remote, this national monument does not attract crowds of visitors. However, it is highly recommended. It is also a good starting point for other trips. Highway 85 ends at the Mexican border (Lukeville on the U.S. side and Sonoita on the Mexican side). From Sonoita, Route 8 leads to Puerto Penasco (see farther on, El Pinacate Natural Park). To the east lies the Papago Indian Reservation. Several primitive roads run through this region of spectacular desert mountains (especially Baboquivari Peak). Here it is possible to explore both the upper and lower Sonoran zones, but it is a good idea to get information about road conditions before heading off.

Arizona: Superstition Mountains and Apache Junction (21)

Take Highway 60/89 east from Phoenix to Apache Junction. From here a visitor can drive about 135 miles (220 km) along a loop road through wild desert. The road goes to Lake Roosevelt, the mining town of Globe, and then back to Apache Junction. This mostly dirt road was built in 1905 to transport supplies from Phoenix to Globe to the workers who were building Roosevelt Dam. The road follows the route of the ancient Apache Indian trail through the canyons of the Salt River. The drive is strenuous and not advisable for the faint-hearted. Some parts of the route (beyond the village of Tortilla Flat) skirt steep precipices. However, the trip is wonderfully spectacular and offers incredible views of high and low desert stretching as far as the eye can see.

Along the trail you can visit Superstition State Park, where you can see the curious Superstition Mountains. This wilderness area is crossed by many trails, and it is a wise idea not to leave them when hiking. The winding canyons and rock formations can cause hikers to lose their sense of direction.

Farther up the Apache trail, you can stop at Fish Creek

Canyon with its enormous multicolored walls, then move on to the Roosevelt Dam which is notable for aquatic birds. Other stops might include Tonto National Monument and the Boyce Thompson Southwestern Arboretum, where there is an excellent collection of cacti and other plants of the Sonoran Desert, in addition to numerous animals.

On the way back to Apache Junction on Route 60/89, a detour south along Route 89 (from Florence Junction) is well worth the while. The road crosses a rich stretch of saguaro forest and lower Sonoran zone desert. Watch for Harris' hawk in this area.

Arizona: Wupatki National Monument (22)

This national monument covers only 13.5 sq. miles (35 sq. km). It can be reached by Highway 89 north from Flagstaff. A side road, clearly marked, loops east and north through Sunset Crater (an extinct volcano, with recent lava flows) and through Wupatki.

This national monument lies on the edge of the Painted Desert, and the scenery is spectacular. Animals are scarce, but some birds, rodents, and collared lizards can be found. The lizards are particularly frequent around Indian ruins.

Arizona: Grand Canyon National Park (23)

The Grand Canyon is undoubtedly the most famous and spectacular national park of the region, as well as the one most often visited. On the south rim of the canyon is an airport, used mainly by small planes that take tourists for morning and evening flights along the main part of the canyon. The park is reached most easily via Route 180 north from Interstate 40 (west of Flagstaff). Complete tourist facilities are located on the south rim.

The Grand Canyon is a huge park that stretches for nearly 70 miles (110 km) and varies from 6 to 25 miles (10 to 40 km) wide. Visitors can make the trip down to the river on mule back or can hike down following several trails. The hike is quite spectacular but strenuous since the terrain is rugged.

A visit to the north rim is also recommended. Take Route 64 east from the south rim to Route 89, then turn north. If starting the trip from Flagstaff, go directly north on Highway 89. Before Page, turn west on Alt 89 to Jacob Lake, then south on Route 67 to the North Rim. The crowds are not as large here, and the forests at higher elevations extend over a huge area. Among the common trees are aspens, ponderosa pines, and junipers. Trails lead down into the canyon, but the hike is long and tough.

Following pages: An impressive natural arch in Arches National Park, Utah, is the result of water and wind erosion, the work of thousands of years.

Arizona:
Kofa National
Wildlife Refuge (24)

This wildlife refuge covers about 880 sq. miles (2,300 sq. km) of desert mountains in southwestern Arizona. The area is an enormous wilderness with no tourist facilities. It is reached by Route 95 north from Interstate 8 (near Yuma), or south from Interstate 10 (east of Blythe, California). The refuge is home to such interesting species as the bighorn sheep and the California fan palm.

MEXICO

In northern Mexico, there are many areas of natural interest. Most of them, though, including the national parks, are difficult to reach except by an organized expedition. Moreover, many of these areas are largely uninhabited, and there are no local facilities. Tour operators, however, have made some areas relatively easy to reach, especially for persons interested in natural history. Among the most frequent excursions are winter boat trips to Scammon's Lagoon in Baja California to watch gray whales and various species of seabirds. Most travel agents can provide details on tours that are available.

El Pinacate
Natural Park (25)

This park is located in the northern part of the state of Sonora and is one of the few Mexican parks near a fairly good road. It is reached along Route 8 from the town of Sonoita on the Arizona border. Route 8 heads south toward Puerto Penasco. The road passes through arid Sonoran Desert, and many sand dunes can be seen. The main park is well to the west of the road, about halfway from Sonoita to Puerto Penasco. There are no tourist facilities, and a long hike requires careful planning. Many species of cacti are located along the road, among which the aguaro and the giant cardon are the most prominent. Among the animals, the peccary, various rattlesnakes, the Gila monster, thrashers (Leoconte's thrasher is almost always present on the dunes close to Puerto Penasco), the cactus wren, and the roadrunner can be seen.

GLOSSARY

abyss a deep fissure or split in the earth; a bottomless pit.

adaptation change or adjustment by which a species or individual improves its condition in relationship to its environment.

algae primitive organisms which resemble plants but do not have true roots, stems, or leaves. Algae are usually found in water or damp places.

amphibian any of a class of vertebrates that usually begin life in the water as tadpoles with gills and later develop lungs.

arid lacking enough water for things to grow; dry and barren.

atmosphere the gaseous mass surrounding the earth. The atmosphere consists of oxygen, nitrogen, and other gases, and extends to a height of about 22,000 miles (35,000 km).

basin all the land drained by a river and its branches.

butte a steep hill standing alone in a plain.

continent one of the principal land masses of the earth. Africa, Antarctica, Asia, Europe, North America, South America, and Australia are regarded as continents.

crest a tuft, ridge, or similar growth on the head of a bird or other animal.

diurnal active during the day.

dominant that species of plant or animal which is most numerous in a community, and which has control over the other organisms in its environment. Dominant species always grow in great numbers.

ecology the relationship between organisms and their environment. The science and study of ecology is extremely important as a means of preserving all the forms of life on earth.

environment the circumstances or conditions of a plant or animal's surroundings. The physical and social conditions of an organism's environment influences its growth and development.

erosion natural processes such as weathering, abrasion, and corrosion, by which material is removed from the earth's surface.

extinction the process of destroying or extinguishing.

fault a fracture or split in a rock mass, accompanied by movement of one part along the split.

fauna the animals of a particular region or period. The fauna of any specific place on earth is determined by the animals' ability to adapt to and thrive in the existing environment conditions.

flora the plants of a specific region or time.

furrow a narrow groove made in the ground.

geology the science dealing with the physical nature and history of the earth. Geology includes a study of the structure and development of the earth's crust, the composition of its interior, individual types of rock, and the forms of life which can be found.

germinate to sprout or cause to sprout or grow.

glaciers gigantic moving sheets of ice that covered great areas of the earth in an earlier time. Glaciers existed primarily in the Pleistocene period, one million years ago.

habitat the area or type of environment in which a person or other organism normally occurs.

humid containing a large amount of water or water vapor.

hydroelectric producing or having to do with the production of electricity by water power or by the friction of water or steam.

inflorescence the arrangement of flowers on a stem.

intrusion the invasion of liquid magma into or between solid rock.

larva the early, immature form of any animal that changes structurally when it becomes an adult.

latitude the angular distance, measured in degrees, north or south of the equator.

mammal any of a large class of warm-blooded, usually hairy vertebrates whose offspring are fed with milk secreted from special glands in the female.

metamorphosis a change in form, shape, structure, or substance as a result of development.

nocturnal referring to animals that are active at night.

omnivore an animal that eats both plants and animals.

organism any individual animal or plant having diverse organs and parts that function as a whole to maintain life and its activities.

photosynthesis the process by which chlorophyll-containing cells in green plants convert sunlight into chemical energy and change inorganic into organic compounds.

physiology the branch of biology dealing with the function and processes of living organisms or their parts and organs.

playa a desert basin that temporarily becomes a shallow lake after heavy rains.

plumage the feathers of a bird. A bird's plumage can provide camouflage, aid in identification, and play an important role in mating rituals.

precipitation water droplets or ice particles condensed from water vapor in the atmosphere, producing rain or snow that falls to the earth's surface.

promontory a peak of high land that juts out into a body of water.

relict a plant or animal species living in isolation in a small local area as a survivor from an earlier period or as a remnant of an almost extinct group.

reptile a cold-blooded vertebrate having lungs, a bony skeleton, and a body covered with scales or horny plates.

riverine on or near the banks of a river.

salinity of or relating to the saltiness of something.

shrub a low, woody plant with several permanent stems instead of a single trunk; a bush.

silt a fine-grained sediment carried and deposited by moving water.

species a distinct kind, sort, variety, or class. Plant and animal species have a high degree of similarity and can generally interbreed only among themselves.

succulent having thick, fleshy tissues for storing water.

symbiosis the living together of two kinds of organisms, especially where such an association provides benefits or advantages for both. Algae and fungi in symbiosis form lichens.

tectonic plate one of several portions of the earth's crust which has resulted from geological shifting. The scientific theory of plate tectonics helps to explain how mountain chains are formed.

terrestrial living on land rather than in water, air, or trees.

transpiration the giving off of moisture through the surface of leaves and other parts of the plant.

tributary a small stream which flows into a larger one.

viviparous bearing or bringing forth living young instead of laying eggs.

INDEX

Agave plants, 31, 79
Air pollution, 7
American Southwest, arid regions, 9
Amphibians, 75-78, 82-84
Antelope ground squirrel, 57-58
Antelope jackrabbit, 67, 70
Anza-Borrego Desert, 104
Aravaipa Canyon, 80, 81
Arches National Park, 119-120
Areas of natural interest
   Arizona, 114-119, 122
   California, 107, 110
   Colorado, 111
   Mexico, 122
   Nevada, 110
   New Mexico, 111
   Texas, 111, 114
   Utah, 108-111, 112-113, 120-121
Asexual reproduction, 31
Aspens, 95
Atacama Desert, 9

Badger, 73-74
Bahia de la Conception, 16-17
Baja California, 10, 11, 15, 16-17, 30, 33, 34
Banner-tailed kangaroo rat, 57, 60
Barrel cacti, 31, 32
Beavertail cactus, 31
Beetles, 39
Big Bend region, 19
Bighorn sheep, 94-95
Birds
   birds of prey, 53-55, 87-89, 93
   high desert, 74-75
   low desert, 51-56
   nighttime birds of prey, 93
   riparian forests, 88-94
   small perching, 54-55
Bison, 73
Black-chinned hummingbird, 92
Black hawk, 87-88
Black phoebe, 87
Black-tailed jackrabbit, 67, 70, 71-72
Black-throated sparrow, 54-55
Black vulture, 88, 89
Black widow, 41, 42-43
Blister beetle, 39
Bluehead sucker, 25
Bony-tailed chub, 25
Boojum, 35
Brewer's sparrow, 75
Brine flies, 27
Broad-billed hummingbird, 92
Brown recluse, 41
Brown vine snake, 85
Buntings, 74, 75
Butterflies, 38, 39, 87
Buttes, 27

Cacti, 10, 27, 29-32
Cactus forest, 19
Cactus wren, 51
California, areas of natural interest, 107, 110
California fan palms, 32, 33
Cantharidin, 39
Canyon de Chelly, 78
Canyonlands National Park, 112-113
Canyons
   amphibians and reptiles of, 82-86
   ecosystems of, 79-80
   fish of, 80-82
   mammals of, 94-104
   vegetation of, 79-80
Canyon tree frog, 82-83
Canyon wren, 89
Cardon cactus, 30
Carnivores, 63, 70, 72-74, 96-104
Catclaw, 33
Cat family, 99-104
Cave Creek Canyon, 79
Centipedes, 41
Cercidium microphyllum palverde species, 33
Chigger, 41
Chihuahuan Desert, 8, 10, 11, 15-16, 18-19, 37
Chipmunks, 95-96
Chipping sparrow, 75
Chisos Mountains, 19
Chubs, 25, 81
Chuckwalla, 44
Cirio plant, 32, 35
Cliff chipmunk, 95-96
Coati, 96-98
Colorado, areas of natural interest, 111
Colorado River, 13, 15, 21-26, 81
Coppery-tailed trogon, 89
Coral snake, 84
Costa's hummingbird, 90-91, 92-93
Cottonwoods, 27, 87
Cougar, 99-103
Coyote, 70, 72
Creeping devil cactus, 32
Curve-bill thrasher, 51

Daces, 81
Dams, 7, 23-25
Death Valley, 12-13, 36-37
Deer, 66-67, 68-69, 95
Dehydration, 37
Desert ecosystems, 6-7
Desert fox, 44
Desert pupfish, 82
Desert shrew, 60-61
Desert tortoise, 44, 45, 49, 51
Desert varnish, 13, 27, 79
Dogwoods, 27
Doves, 51-52
Dry-period strategies, flowering plants, 37
Dunes area, Nevada, 11

Ecosystems
   canyons and riverbanks, 79-80
   desert, 6-7

Elephant trees, 33-34
Elf owl, 53, 54, 55
El Gran Desierto of Sonona, 11, 15
Erosion, 13-14, 15, 21, 22, 27
Estivation, 58

False coral snake, 84
Ferruginous owl, 93
Firs, 26
Fish, 25, 80-82
Flammulated owl, 93
Flannelmouth sucker, 25
Flowering agave, 31
Flowering plants, dry period
  strategies of, 37
Flycatchers, 54-55, 87
Fossil water, 7
Foxes, 44, 63
Freshwater springs, 81
Fringe-toed lizard, 44, 45
Frogs, 75-77, 82-84

Gambel's quail, 52, 54, 55
Giant cardon cactus, 31
Gila cypha fish species, 25
Gila monster, 46-47
Gila sucker, 81
Gila woodpecker, 50, 51, 53
Gilded flicker, 54-55
Glen Canyon Dam, 23-25
Grand Canyon, 21-26
Gray Arizona squirrel, 95, 96
Gray hawk, 88-89
Great horned owl, 53-55
Great Salt Lake, 27, 108-109
Ground squirrels, 77
Gulf of California, 22

Hackberries, 27
Harlequin quail, 87
Hawks, 87
Heloderma genus, lizards, 46
High deserts
  amphibians of, 75-78
  birds of, 74-75
  carnivores of, 70, 72-74
  described, 19-20
  hoofed animals and rabbits of,
    64-72
Horned lizard, 44, 45-46
Horse latitudes, 9
House finch, 55
Hummingbirds, 87, 89-93
Hurricanes, 15

Insects, 39, 41, 44
Invertebrate predators, 39-44
Invertebrates, 29, 81

Jackrabbits, 67, 70, 71-72
Jaguar, 99-100, 101
Jaguarundi, 103
Javilina, 61

Joshua Tree National Monument, 34
Joshua trees, 34-35
Jumping cholla cactus, 31-32
Juniper berries, 79
Junipers, 10, 27, 79, 89, 99

Kangaroo rats, 57
Kit fox, 63

Lake Powell, 25
Lark bunting, 75
Legume family, 33
Leopard lizard, 44, 45
Lizards, 44-47
Loach minnow, 81
Longfin dace, 81, 82
Low deserts
  birds of, 51-56
  carnivores of, 63
  described, 11-15
  rodents of, 57-61
  small animals of, 38-50
  vegetation of, 29-38
Lynx, 102-103

Madera Canyon, 79
Mammals
  canyon, 95-104
  small, 57-61
Mammillaria cacti, 31
Marine blue butterfly, 38
Meadowlark, 74
Mesquite, 32, 33
Mescal, 79
Mexico, areas of natural interest, 122
Millipedes, 41
Mockingbird, 53
Mogollon Rim, 19, 20, 21
Mojave Desert, 10, 11, 13
Monarch butterfly, 38, 39
Monument Valley, 6-7, 26-27
Mormon tea, 27
Mountain lion, 100-103
Mule deer, 66-67, 95
Myriapods, 39, 41

Namib Desert, 9
Nevada, areas of natural interest, 110
New Mexico, areas of
  natural interest, 111
Nighttime birds of prey, 93
North American plate, 13

Oak Creek Canyon, 20-21
Oaks, 10, 27, 79, 87, 89, 95, 99
Oases, 81
Ocelot, 103
Ocotillo plant, 28, 31, 35, 37
Opuntia genus, cacti, 31-32
Organ pipe cactus, 30
Organ Pipe Cactus National
  Monument, 30, 81, 82
Owls, 53-55, 93

Pacific plate, 13
Pack rat, 59-60
Painted Desert, 20-21
Pajarito Mountains, 85
Paloverde, 32, 33
Papago Indians, 33, 80
Peccary, 61-63
Phainopepla, 55
Photosynthesis, 33, 37
Pinacate beetle, 38
Pines, 26, 27
Pinicate Desert, 15
Pink brine shrimp, 27
Pinnate leaves, 33
Pinyon pine, 27
Pit viper family, 47
Predators
  birds of prey, 53-55, 87-89, 93
  canyons, 96-100
  high desert, 70, 72-74
  invertebrate, 39-44
  low desert, 63
  snakes, 47-49, 84-85
Prickly pear cactus, 31
Pronghorn, 64-66

Quails, 52, 54, 55, 87
Quaking aspen, 26
Quitobaquito Springs, 82

Rain shadows, 9
Ramsay Canyon, 79
Rattlesnakes, 47-49, 84
Ravens, 74
Razorback sucker, 25
Reptiles, 44-49, 84-85
Ringtail, 98, 99
Riparian forests, 10, 79, 80
  birds of, 88-94
Riverbank ecosystems, 79-80
Rivoli's hummingbird, 92, 93
Roadrunners, 52, 53
Rodents, 57-61
Round-tail chub, 25, 81
Round-tail ground squirrel, 58

Sagebrush, 27
Saguaro cactus, 29-30, 31
Sahara, 9
Salt flats, 27
Salton Sea, 23, 82
Savanna sparrow, 75
Scorpions, 41
Scott's oriole, 89
Screech owl, 93
Sea of Cortez, 15
Seas of sand, 10-11
Senita cactus, 30
Shrubs, 10, 25, 33-38
Sidewinder rattlesnake, 49
Sierra Madre ranges, 18-19, 82
Silting, Colorado River, 23, 25
Sinks of water, 80

Small carnivores, 96-99
Small margay, 103
Small perching birds, 54-55
Snakes, 47-49, 84-85
Snout butterfly, 38
Sonora chub, 81
Sonoran Desert, 10, 11, 13, 15, 19, 29, 33, 34, 80, 81
Sonoran green toad, 76-77
Sonoran sucker, 81
Spadefoot toad, 76-77
Sparrows, 54-55, 74-75
Speckled dace, 25
Spiders, 40-41, 42-43
Spike dace, 81
Spruce, 26
Squawfish, 25
Squirrels, 57-58, 77, 95-96
Strip mining, 7
Suckers, 25
Sulphur-bellied flycatcher, 87
Sulphur butterfly, 38
Sycamore Canyon, 79
Sycamores, 87, 95

Tamrisk shrub, 25
Tarahumara frog, 82, 84
Tarantula, 40
Tarantula hawk wasp, 40-41
Teddy-bear cholla cactus, 31
Texas,
 areas of natural interest, 111, 114
Thrashers, 51
Toads, 76-77
Tree frogs, 75-76
Turkey vulture, 88, 89

Utah, areas of natural interest, 108-111, 112-113, 120-121

Vegetation
 canyon, 79-80
 low deserts, 29-38
Vertebrates, 28
Volcanic cones, 14-15
Volcanic intrusions, 14
Vultures, 88, 89

Wasps, 40

Water erosion, 21, 22, 27
Water table, 29
Wetlands, 81
Whisker cactus, 30
Whiskered owl, 93
White-collared raven, 74
White-crowned sparrow, 75
White-necked raven, 74
White Sands region,
 New Mexico, 11, 18-19
White-tailed deer, 66-67, 68-69, 95
White-throated wood rat, 60
White-winged dove, 54-55
Wind erosion, 13-14, 22, 27
Woodpeckers, 50, 51, 52-53
Wood rat, 59-60
Wrens, 51, 89

Yucca moth, 35
Yucca trees, 34-35, 89

Zebra-tailed lizard, 44
Zone-tailed hawk, 89

REFERENCE -- NOT TO BE
TAKEN FROM THIS ROOM

DATE DUE

DATE DUE

```
574.5    Wingfield, John C
Win
         Deserts of America
```

15.00

25598

Laramie Junior High
1355 N. 22nd
Laramie, WY 82070